THE FRONTIERS COLLECTION

Series editors

Avshalom C. Elitzur
Unit of Interdisciplinary Studies, Bar-Ilan University, 52900 Ramat-Gan, Israel
e-mail: avshalom.elitzur@weizmann.ac.il

Laura Mersini-Houghton
Department of Physics, University of North Carolina, Chapel Hill,
NC 27599-3255, USA
e-mail: mersini@physics.unc.edu

T. Padmanabhan
Inter University Centre for Astronomy and Astrophysics (IUCAA), Pune, India

Maximilian Schlosshauer
Department of Physics, University of Portland, Portland, OR 97203, USA
e-mail: schlossh@up.edu

Mark P. Silverman
Department of Physics, Trinity College, Hartford, CT 06106, USA
e-mail: mark.silverman@trincoll.edu

Jack A. Tuszynski
Department of Physics, University of Alberta, Edmonton, AB T6G 1Z2, Canada
e-mail: jtus@phys.ualberta.ca

Rüdiger Vaas
Center for Philosophy and Foundations of Science, University of Giessen,
35394 Giessen, Germany
e-mail: ruediger.vaas@t-online.de

T0210642

THE FRONTIERS COLLECTION

The books in this collection are devoted to challenging and open problems at the forefront of modern science, including related philosophical debates. In contrast to typical research monographs, however, they strive to present their topics in a manner accessible also to scientifically literate non-specialists wishing to gain insight into the deeper implications and fascinating questions involved. Taken as a whole, the series reflects the need for a fundamental and interdisciplinary approach to modern science. Furthermore, it is intended to encourage active scientists in all areas to ponder over important and perhaps controversial issues beyond their own speciality. Extending from quantum physics and relativity to entropy, consciousness and complex systems—the Frontiers Collection will inspire readers to push back the frontiers of their own knowledge.

More information about this series at http://www.springer.com/series/5342

For a full list of published titles, please see back of book or springer.com/series/5342

Anthony Aguirre · Brendan Foster
Zeeya Merali
Editors

HOW SHOULD HUMANITY STEER THE FUTURE?

FQXi
FOUNDATIONAL QUESTIONS INSTITUTE

Springer

Editors
Anthony Aguirre
Department of Physics
University of California
Santa Cruz, CA
USA

Zeeya Merali
Foundational Questions Institute
New York, NY
USA

Brendan Foster
Foundational Questions Institute
New York, NY
USA

ISSN 1612-3018 ISSN 2197-6619 (electronic)
THE FRONTIERS COLLECTION
ISBN 978-3-319-37340-9 ISBN 978-3-319-20717-9 (eBook)
DOI 10.1007/978-3-319-20717-9

Springer Cham Heidelberg New York Dordrecht London
© Springer International Publishing Switzerland 2016
Softcover re-print of the Hardcover 1st edition 2016

Printed on acid-free paper

Springer International Publishing AG Switzerland is part of Springer Science+Business Media
(www.springer.com)

Preface

This book is a collaborative project between Springer and The Foundational Questions Institute (FQXi). In keeping with both the tradition of Springer's Frontiers Collection and the mission of FQXi, it provides stimulating insights into a frontier area of science, while remaining accessible enough to benefit a non-specialist audience.

FQXi is an independent, nonprofit organization that was founded in 2006. It aims to catalyze, support, and disseminate research on questions at the foundations of physics and cosmology.

The central aim of FQXi is to fund and inspire research and innovation that is integral to a deep understanding of reality, but which may not be readily supported by conventional funding sources. Historically, physics and cosmology have offered a scientific framework for comprehending the core of reality. Many giants of modern science—such as Einstein, Bohr, Schrödinger, and Heisenberg—were also passionately concerned with, and inspired by, deep philosophical nuances of the novel notions of reality they were exploring. Yet, such questions are often overlooked by traditional funding agencies.

Often, grant-making and research organizations institutionalize a pragmatic approach, primarily funding incremental investigations that use known methods and familiar conceptual frameworks, rather than the uncertain and often interdisciplinary methods required to develop and comprehend prospective revolutions in physics and cosmology. As a result, even eminent scientists can struggle to secure funding for some of the questions they find most engaging, while younger thinkers find little support, freedom, or career possibilities unless they hew to such strictures.

FQXi views foundational questions not as pointless speculation or misguided effort, but as critical and essential inquiry of relevance to us all. The Institute is dedicated to redressing these shortcomings by creating a vibrant, worldwide community of scientists, top thinkers and outreach specialists who tackle deep questions in physics, cosmology, and related fields. FQXi is also committed to engaging with the public and communicating the implications of this foundational research for the growth of human understanding.

As part of this endeavor, FQXi organizes an annual essay contest, which is open to everyone, from professional researchers to members of the public. These contests are designed to focus minds and efforts on deep questions that could have a profound impact across multiple disciplines. The contest winners are chosen by a combination of input from entrants, FQXi Members, and a panel of judges, and up to twenty prizes are awarded. Each year, the contest features well over a hundred entries, stimulating ongoing online discussion long after the close of the contest.

We are delighted to share this collection, inspired by the 2014 contest, "How Should Humanity Steer the Future?" In line with our desire to bring foundational questions to the widest possible audience, the entries, in their original form, were written in a style that was suitable for the general public. In this book, which is aimed at an interdisciplinary scientific audience, the authors have been invited to expand upon their original essays and include technical details and discussion that may enhance their essays for a more professional readership, while remaining accessible to nonspecialists in their field.

FQXi would like to thank our contest partners: Jaan Tallin, The Gruber Foundation, The John Templeton Foundation, and Scientific American. The editors are indebted to FQXi's scientific director, Max Tegmark, and managing director, Kavita Rajanna, who were instrumental in the development of the contest. We are also grateful to Angela Lahee at Springer for her guidance and support in driving this project forward.

2015   Anthony Aguirre
                                                                               Brendan Foster
                                                                               Zeeya Merali
                                                                               www.fqxi.org

Contents

Chapter 1
Introduction

Anthony Aguirre, Brendan Foster and Zeeya Merali

> *Science fiction writers foresee the inevitable, and although problems and catastrophes may be inevitable, solutions are not.*
> Isaac Asimov (1931) [1]

> *We are in danger of destroying ourselves by our greed and stupidity. We cannot remain looking inwards at ourselves on a small and increasingly polluted and overcrowded planet.*
> Stephen Hawking (2010) [2]

Improving the future for our civilization is one of the foremost goals of both the sciences and the humanities. These endeavours allow us to learn from both our past mistakes and successes, to anticipate potential catastrophes, and to develop technologies and lines of thinking to preempt them. Yet dystopic visions of the future—often based on the unchecked rise of the very scientific and technological innovations designed to help society—abound in literature and film, while optimistic ones are more rare.

In 2014, FQXi launched an essay contest with the aim of redressing the balance by encouraging entrants to think about ways to avoid potentially self-fulfilling prophecies of doom and gloom. "How," we asked, "should humanity steer the future?"

This was one of the broadest questions that we had yet posed for an essay contest, and required participants to not only imagine future pitfalls, but also to outline

A. Aguirre (✉)
Department of Physics, University of California, Santa Cruz, CA, USA
e-mail: aguirre@scipp.ucsc.edu

B. Foster · Z. Merali
Foundational Questions Institute, New York, NY, USA
e-mail: foster@fqxi.org

Z. Merali
e-mail: merali@fqxi.org

© Springer International Publishing Switzerland 2016
A. Aguirre et al. (eds.), *How Should Humanity Steer the Future?*,
The Frontiers Collection, DOI 10.1007/978-3-319-20717-9_1

practical strategies to mitigate them. Our ever-deepening understanding of physics has enabled technologies and ways of thinking about our place in the world that have dramatically transformed humanity, and the world that we live in, over the past several hundred years. Some of the resulting problems that will face future generations are already apparent. It will require global efforts to address human-induced climate change, for instance. Yet, as we have seen, it is often difficult to persuade governments and the public to establish policies and habits now that may only reap benefits over the longterm. Other threats to humanity that could arise from future technology, such as artificial intelligence, have barely even entered serious public discussion. Many others will take an unknown form that we have yet to imagine based on the radically different modes of thought and fundamentally new technologies that could become relevant in the coming decades.

In this vein, we asked participants to consider what they believe the best state that humanity could realistically achieve might be, what plan would be needed to reach that point, and who would need to implement that plan. The contest drew 155 entries from thinkers both within and outside the academic system. It proved a resounding success, raising many new lines of inquiry and demonstrating the same creativity, big-picture thinking and depth of understanding seen in previous essay contests. This success, and the urgency of many of the issues brought to light, inspired the foundation of a separate body, the Future of Life Institute (http://futureoflife.org/), which supports initiatives for safeguarding life and developing optimistic visions of the future. FLI has subsequently grown rapidly, with the successful launch of several initiatives addressing the future promise and perils of artificial intelligence.

This volume brings together the top 14 prize-winning entries from the contest. Some identify particular risks to humanity's security, while others propose general changes that could be made now to education and research in order to arm society against threats of any form—whether natural, human-induced, or even from alien civilizations. Still others address how to make society receptive to any proposed changes.

Our first-prize winner, Sabine Hossenfelder, challenges the value of the essay question itself. In Chap. 2, she notes that even if we knew how best to steer the future of humanity, that knowledge will be of little use if the wider population does not enforce it. She sketches a strategy for disseminating insights to the public in a palatable manner to maximise their impact, enabling people to evaluate possible courses of action for themselves in an informed manner.

The next two chapters deal with assessing the specific form of longterm risks. Given how difficult it is to predict tomorrow's weather, in Chap. 3, Tommaso Bolognesi, considers how best to accurately simulate far-future scenarios for humanity's fate based on existing data sets. In Chap. 4, Daniel Dewey calls for governments to invest in research into possible threats arising from biological engineering and artificial intelligence.

Chapters 5–8 offer strategies to arm humanity against catastrophes, whatever they may be, by changing people's attitudes about their influence over the future. Preston Estep III and Alexander Hoekstra, in Chap. 5, advocate focusing on techniques for strengthening the human mind. Dean Rickles argues that people often underestimate their ability to affect the future. In Chap. 6, he suggests that inspiration to rectify this could come from interpretations of quantum mechanics that highlight the role of the observer on measurements and from the philosophy of time. Rick Searle similarly discusses how the impact of technology on society has led to a lost sense of freedom over the future. In Chap. 7, he argues that this could be remedied by re-establishing the "Utopian ideal". By contrast, Tejinder Singh makes the case that fixating on the past or obsessing anxiously over the future can have a negative effect on the mind. Instead, he advises that humanity should learn to live in the here and now. Enlightenment, he says in Chap. 8, is not for the Buddha alone.

A number of winners proposed ways to improve current science education and research. In Chap. 9, Travis Norsen argues that science teaching should be less dogmatic, with more emphasis on the historical development of ideas and scientific controversies, so that scientists are better equipped to deal with contentious issues in the future. Jonathan Dickau values the role of play in learning and physics research and, in Chap. 10, he argues that recognising this will fuel innovation.

In Chap. 11, Mohammed Khalil proposes a number of changes within the academic system, including developing new specialisations at undergraduate level to deal specifically with energy solutions, encouraging collaboration between various disciplines, enhancing public understanding of science through online courses, and using *Wikipedia* as a model for generating online review articles summarising new research. The issue of how to store information over the longterm is also addressed, in Chap. 12, by Jens Niemeyer, who notes that as ever-increasing amounts of data are held in digital form, the risk of losing vast tracts of knowledge in a global disaster is also raised. He argues that a secure physical repository is needed to protect humanity's heritage.

Chapters 13–15 look beyond Earth when considering global security. In Chap. 13, George Gantz invites people to draw on humanity's most positive values such as love, respect, and humility, to prepare them for possible first contact with an alien civilization. Flavio Mercati uses lessons from history to make the case that humanity needs to achieve equilibrium with its environment. In Chap. 14, he considers future scenarios in which people will need to terraform and colonise other planets and argues that preserving biodiversity will be essential for their success. Chapter 15 closes the volume with a novel work of fiction by Georgina Parry that imagines a highly technological world in the wake of overpopulation and climate change. She explores the issues surrounding the society of survivors as they contemplate migration to another world.

This compilation brings together the most diverse range of winners of any of FQXi's essay contests. The contributors to this volume include academic researchers from the fields of high energy physics, theoretical and computational cosmology, philosophy and quantum gravity, and those who now work, or have worked previously, in genetics, aspects of mathematical modelling, software engineering, audio and video production, business, and science education. This mix is appropriate given the broad scope of the question that FQXi posed and its widescale potential impact. It serves to highlight the importance of interdisciplinary collaboration when considering the longterm future of our civilization.

References

1. Asimov, I.: How Easy to See the Future, Natural History magazine (April 1975); later published in Asimov on Science Fiction (1981)
2. Hawking, S.: Interview with Larry King, CNN, 10 September 2010

Chapter 2
How to Save the World

In Five Simple Steps

Sabine Hossenfelder

Abstract If you knew how humanity should steer the future, what difference would it make? The major challenge that humanity faces today is not that we lack ideas for what to do, as I am sure this essay contest will document. No, the major challenge, the mother of all problems, is to convert these ideas into courses of action. We fail to act in the face of global problems because we do not have an intuitive grasp on the consequences of collective human behavior, are prone to cognitive biases, and easily overwhelmed by data. We are also lazy and if intuition fails us, inertia takes over. How many people will read these brilliant essays? For the individual, evaluating possible courses of action to address interrelated problems in highly connected social, economic and ecological networks is presently too costly. The necessary information may exist, even be accessible, but it is too expensive in terms of time and energy. To steer the future, information about our dynamical and multi-layered networks has to become cheap and almost effortless to use. Only then, when we can make informed decisions by feeling rather than thinking, will we be able to act and respond to the challenges we face.

2.1 The Problem

The most remarkable fact about humans is the utter uselessness of our infants. Humans, in contrast to all other species, must learn almost everything necessary for survival. It takes us a long time to reach maturity, time in which parents have to prevent their offspring from eating sand, chopping off fingers, or accidentally wiping out the human race by growing super-resistant bugs behind their ears.

But our ability to learn, combined with technics to communicate information, is also what enables us to adapt to changing environments faster than gene selection could possibly achieve. We are awed by inborn knowledge—butterflies that recall routes of their ancestors—but we outpaced our competitors by changing the rules of what it means "to adapt" itself. We do not wait for physiological changes to result

S. Hossenfelder (✉)
Nordita, Roslagstullsbacken 23, 10691 Stockholm, Sweden
e-mail: sabine.hossenfelder@gmail.com

© Springer International Publishing Switzerland 2016
A. Aguirre et al. (eds.), *How Should Humanity Steer the Future?*,
The Frontiers Collection, DOI 10.1007/978-3-319-20717-9_2

in better Darwinian fitness. Instead, we modify our environment, ourselves, and our interaction with the environment to get there faster, to "fit" better than any other species.

The root of the problems that humanity faces today is that our adaptation as a species has fallen behind the changes we have induced ourselves. As human interactions get more complex, as networks spread globally and become tightly coupled, we need systems that are able to learn and in return help let us learn about the system. But we don't have them.

The political, economic, and social systems that govern our lives are presently adaptive by trial and error. But much like gene selection is too slow to have yet adapted humans to a mostly sedentary city life and goat memes, the adaptation of our systems by trial and error is too slow to solve the problems we presently face. May that be climate change, the global water crisis, the big garbage swirl, or the fragility of our financial systems—our inability to process these problems means that we, as the actors in the system, do not respond, indeed cannot respond, to the information we have in a timely manner. And so, problems persist and build up.

The necessary information for individuals to learn and react to systemic trends may be available, even accessible, but it is too expensive. Information is presently costly, not necessarily financially, but in the amount of effort required to obtain and understand it. Relevant information is too difficult to find or comprehend and doing so requires too much time and energy. Blaming people for being politically disinterested, scientifically illiterate, or plainly unintellectual doesn't do anything to address the costliness of information and thus doesn't do justice to the origin of the problem. The individual investment necessary to process information about trends and relations in our systems is currently too high and personal benefits do not outweigh the disadvantages.

There is no shortage of well-intentioned institutions and organizations that aim at one or several of humanity's problems. The biggest, most existential, problems have been collected by the Future of Humanity Institute [1]. These are the problems that can lead to extinction, near-extinction or progress stagnation of the human species, including but not limited to nuclear accidents, asteroid impacts, artificial intelligence and biotechnology. In case you're not a worrier, check their webpage.

The Future of Humanity Institute is also a point in case because it has next to no influence on the future of humanity. The same can be said about all other initiatives that collect data about our global networks and use buzzwords like complexity and interdisciplinarity. For example the United Nations Global Pulse [2], the FutureICT [3] or the PSIR Model [4]. All such initiatives fail to address the main problem, which isn't to collect information, but feeding this information back into the system, back to the many humans who are the initiators of change.

Some believe quantum computing will solve our problems. It won't. Computation alone cannot solve our problems for the same reason political utopias, however beautiful and ingenious, never solved any problems: Because humans don't care what somebody or some thing thinks they should be doing. They'll do whatever they please. The only way to change their ways is to please them. Please them differently than before, and change will follow.

No, the biggest challenge mankind faces today is not the development of some breakthrough technology. The biggest challenge is to create a society whose institutions integrate the knowledge that must precede any breakthrough technology: The knowledge about the systems themselves that is necessary for the realization, adaptation and use of technologies. All of our big problems today speak of our failure, not to envision solutions, but to turn our ideas and knowledge into reality. We have a social problem, not a technological one.

We reached this gridlock because the human brain did not evolve to understand the consequences of individual actions in networks of billions of people. We are bad in making good long-term decisions and do not care much what happens in other parts of the planet to people we have not and will most likely never meet. We have no intuitive grasp on the collective behavior of large groups and their impact on our environment, and what little grasp we have is prone to cognitive biases and statistical errors, many of which are now subject of new scientific areas like game theory, behavioral economics and decision science.

These cognitive shortcomings are not only obstacles to solving our problems, they *are* the problem. But these are obstacles that science can overcome.

2.2 The Solution

The human brain has the capacity to evaluate decisions that have long-term and large-scale consequences. However, frequently decisions which are beneficial on long time- or distance scales conflict with those on short time- or distance scales. Due to evolutionary developed reward circuits, this conflict is often resolved in favor of short times and distances.[1]

But we know how to solve these problems. We solve them by bringing close that what is far away. This is why people in weight-loss programs (distant) are encouraged to reward themselves (close) for holding onto their diet. This is why they pin photos (distant) on their fridge (close). This is why the World Wildlife Fund lets you adopt baby animals of endangered species (distant) and sends you a certificate (close). This is why you are shown all the photos of hungry, ill, injured or otherwise suffering children. You get the picture. It brings distant information closer and taps onto your emotional responses, which is a fast, simple, and effective reaction. It wires back into the circuits that your brain is used to work with. This wiring can be abused, all right. But used the right way, it carries the solution to our problem.

"Gamification" is a recent variant of this mechanism. Gamification is growing popular to help people balance their own priorities, typically by providing instant rewards (in terms of collecting points) for behavior users previously themselves identified as desirable (say, eating healthy). Seen from a system's perspective, this is an external feedback loop that allows humans to use old brain circuits to adapt good

---

[1] Here, with "distance" I am referring not necessarily to spatial distance, but to distance in social networks and other infrastructure networks.

(here: healthy) behavior faster than gene selection could achieve with a turnover rate of many generations. The interesting aspect about gamification is how little is necessary to make this feedback loop work. All it takes is a simple and intuitive visualization that lets users immediately grasp how well an action matches with their stated goals. The keywords here are: Simple, intuitive, and immediate. This is cheap information.

The solution to our problems is a generalization of this feedback loop: To give people access to cheap information about the consequences of collective human actions, and in return use their reaction to this information to improve the system, i.e. the way individual actions are coordinated.

The point here is not to manipulate people into changing their ways because I or you or some supercomputer thinks it would be better if we'd do more of this or more of that. The point is to help people make decisions. The way we presently make decisions, part of our priorities remain neglected because we cannot assess how well we would be working towards them. It's too complicated, too costly. But it's not like we are happy with this. Most people notice the tension, the neglect of some of their priorities, and are left with bad consciousness, the nagging voice that says you should make better decisions. If only you had the time and it wasn't so difficult.

The feedback system that we need has to give the user an intuitive feeling for how well a decision matches with recorded priorities. If such a feedback in the future can be given by a brain implant, it will be like an additional sense. How does this decision taste? How well does it match with my preferences? Does this choice look harmonious? Does it sound good? Such a feedback is the natural extension of our ability to judge the result of our actions in small groups. This is what it takes to make information cheap, really cheap, so that using it becomes almost effortless.

This feedback loop might include for example information about how well buying a product matches with the relevance one has assigned to certain health goals or its environmental impact or its contribution to the local economy. This is information which a customer doesn't normally have when making a purchase (though economic theory maintains it is taken into account). And even if they had the information, they probably wouldn't study it.

Other examples are questions like: If I dispose of that plastic bottle here, how likely is it to be recycled or to end up in the ocean? If I buy the fair trade coffee, does it work towards something I value? Do I help the homeless guy more by giving him some dollars or by donating that money somewhere? How much of the tax I pay on this item subsidizes projects I support? It's not that people do not care. It's just that in practice it takes too much effort to look into the details. And they actually do not want to know the details. All they want to know is whether, according to best present knowledge, a certain decision works towards their goals. And most of the time that is really all they need to know.

Let me use another example, a somewhat shocking one that however illustrates well distance among people. A recent study by researchers from Princeton University asked participants to judge the competence of political candidates by split-second looks at photos. It turned out that this snap judgment predicted very well who would eventually be voted [5].

How incredibly shallow we are. But forgive us. We decode human faces constantly and effortlessly and the human brain always tries to save energy. We use emotional response to somebody's look to assess how much we can trust them. That's not an optimal assessment for informed decision making. It certainly gives me to think that my opinion of political candidates probably depends on the shape of their nose. I really should go and read all these programs, comments and opinion pieces. But I have an essay to finish before the deadline, then write this overdue report and hurry to pick up the kids from daycare. Maybe I'll look up these candidates next week. Or the week after that. If only information wasn't so costly. If only it wouldn't take up so much time and energy.

But now imagine you could look at a candidate and in fact get a simple, fast, sensorial or emotional feedback how certain selected priorities and interest of this person match with yours. This would dramatically lower the cost of information. It would bring close that what is far away.

The ingredients for closing this feedback loop already exist, they just aren't combined suitably. Above I mentioned gamification to bridge long time distances. Other applications that make information less costly in terms of time and energy are sites dedicated to help you decide which party to vote based on answers to a set of questions, or dating sites that match your interests with potential mates. It's the same mechanism, but too scattered and not broad enough. The more dispersed these applications are, the more effort it takes to use them and the more costly the information becomes. We need it all in one place.

Concretely the feedback loop would work like this:

1. A user creates a personal priority map. In the future this may be done by a brain scan or by analyzing information transmitted from neural implants. Presently, questionnaires and other records must stand in. The questionnaires would cover for example personal values, various aspects of health and social life, political attitudes and personal taste. This should also include users' tolerance for risk and uncertainty because this is relevant to assess how good a match will be. This priority map is personal data that the user can update and expand, and share or make public selected parts of it.
2. Institutions that gather knowledge about the system (statistics, trends, predictions) make it available to users as correlations between actions and individual priorities. In return they use the shared parts of users' maps to obtain better information about the system, notably tensions that arise when priorities conflict, which can indicate problems with the current organization of the system.
3. Whenever a user takes a decision whose impact is likely to exceed the natural human ability to foresee consequences of individual actions he or she consults the priority map. The user can then tell how well a decision matches their recorded priorities and take this into account without having to bother with the details in every single case. The decisions serve to adapt the system.

This consultation of the priority map thus remedies the lack of intuition humans have for the behavior of highly connected and dynamic social networks. The goal is that users are able to make informed decisions with snap judgments: Simple, intuitive, immediate. Only then will we change the ways of the bulk of people on the planet.

In the future, information about matches with personal priorities may be delivered wirelessly to brain implants, constituting an upgrade of humanity for global interactions. I only discussed here the evaluation of selected decisions, not how to find the best possible course of action according to certain criteria. The latter is a much harder problem. We can note in the passing though that it constitutes an optimization problem and thus lends itself for adiabatic quantum computing.

With presently existing technology we have to settle for visualizing a match or mismatch rather than feeling it. The visualization of big data sets and the possibility to manipulate them interactively is rapidly improving, and such interaction with data will already serve to make information dramatically cheaper. And it really has to become cheap.

We do not get anywhere with bemoaning that most people do not understand climate models or do not read information brochures about genetically modified crops. It is time to wake up. We've tried long enough to educate them. It doesn't work. The idea of the educated and well-informed citizen is an utopia. It doesn't work because education doesn't please people. They don't like to think. It is too costly and it's not the information they want. What they want is to know how much an estimated risk conflict with their priorities, how much an estimated benefit agrees with their values. They tolerate risk and uncertainty, but they don't tolerate science lectures. If a webpage takes more than 3 s to load it'll lose 40 % of visitors. Split-second looks at photos. That's the realistic attention span. That's what we have to work with.

In some regards we are already on the way to close this feedback loop. Many scientific institutions share information and take science communication seriously. However, presently this information is still too much, too unclear, and not available to individuals at the right moment.

In other regards we have a long way to go. We do not presently use people's priorities in any systematic way to discover shortcomings in the system and improve it. The economic system to some extent does what we want. After all, it's not like we've been total losers at steering the future of humanity. But the standard theory of the economic system assumes that consumers have full access to relevant information, that they take it into account, and that their decisions reveal their true preferences. However, monetary value is a one-dimensional measure that inevitably disregards the multi-valued reasons people have to invest money, and this projection on a one-dimensional scale means that information is lost.

Concretely, imagine how much more useful book reviews would be if you knew the reviewers' priorities compared to yours, if you knew what they consider a "good book". Imagine how much more useful sales numbers would be if companies knew how important economic and social engagement are for their customers. The economic system alone doesn't give us this information.

Moreover, emotions can capture problems that do not result in actions at all (are not "revealed"). Take the 2008 mortgage crisis as an example. If you read reports from back then, many people clearly felt something was wrong. "Something about that feels very wrong," a banker said, "It makes me sick to my stomach the kind of loans that we do," a mortgage broker was quoted with [6]. But these feelings didn't register in the system. Imagine we could have measured the tension in priorities between, say, keeping their job and acting morally right. This could have been an early warning sign. How many warning signs do we currently miss?

So, what we need for humans to interact intelligently on global scales is a simple and intuitive way for them to tell how much their priorities—long-term as well as short-term, locally as well as globally—match with decisions they can take. That might strike you as a very abstract idea. Let me tell you then where to start making it reality.

2.3 Science Matters

Making information cheap does not make it correct, and as they say: Garbage in, garbage out. Information about the system is only useful to steer the future if it's accurate. There will generically be several information providers for the same correlation and their findings might disagree. This temporary disagreement is in the nature of research which brings us to the process of knowledge discovery and to the system that it operates with.

In the following I refer to the process of knowledge discovery as being executed by the academic system, by which I mean scientific research that is not conducted for profit. Scientific research is of course also conducted for profit, but I will not discuss this here because mixing in the economic system makes things more complicated without making them more insightful.

The academic system plays a pivotal role for establishing the feedback loop that allows our systems to integrate and process globally dispersed information. If we can close the feedback loop and the system can learn, all other problems are self-correcting. Thanks to the scientific method, we need not be afraid of conflicting information and uncertainties. These improve over the course of time, provided the process of knowledge discovery works as desired. Unfortunately, it presently doesn't. The reason is that the academic system too isn't able to learn. But do not despair, because we already know what to do. We need priority maps for scientists.

Personnel in administrative academic positions and scientists are faced today with many complex decision tasks. Everybody agrees that personal assessment of research projects and researchers is the best possible judgment. But not everybody can possibly assess everything and everybody. That is the scientists' problem of too costly information—it would take too much time to read all these papers. There is also other information about the system that scientists do not readily have access to. How good, for example, is the reputation of some university in a country you're not even sure where to find on a world map? Is this research area blooming or in decline? Is that a typical number of coauthors and publications in this field? How do I judge the enthusiasm of this referee?

Because of pure need scientists use whatever means are available to select the relevant pieces of information. Typical problems they face are finding relevant new publications or the most interesting candidates on lists with hundreds of applicants. Presently, they rely heavily on personal connections and some existing measures of scientific success. That is not good for many reasons. Personal connections are inevitably subjective and existing measures are too rough, too inflexible and also too streamlined.

Imagine how much a priority map could help scientists and how much time they would save. Their personal map would include their own research topics and judgment of these, interests in other fields of research and assessments of these, how relevant they believe certain traits to be for the scientific success of candidates, how relevant they think it is that a candidate's research topic matches the local research, and so on. You can add your wishlist here.

The information providers would be scientometric measures that indicate how research areas grow, co-authorship networks, co-citation links between research areas, and other existing measures about the connectivity and impact of scientific research.

These priority maps become truly useful when many people use and (partially) share them, for example their assessment of research results or students. But importantly the maps would be useful already for single users because much relevant information is publicly available. It is for example technically feasible to extract keywords from candidates' publication lists (or research plans) and match them with that of members of a committee. This would be a much better way to find potentially interesting candidates than looking for familiar names. This is just one concrete example of which I could list many, but this isn't the place for lots of details so let me return to the big picture.

The main point is that priority maps have the power to make a difference because they are designed to be useful for scientists. This is in contrast to existing measures for scientific success that are designed to be useful for administrational and external uses and are thus widely met with rejection and cynicism by researchers. The existing measures are used just because they are available and wide-spread, not necessarily because scientists think they are good. With the pressure of having to make a decision, saving time and energy scores higher than other, more idealistic, values. This conflict between individual short-term benefits and collective long-term benefits that turn into individual long-term benefits is the same tension we already discussed earlier. The root of the problem is the same: An accurate assessment of individual consequences from collective trends is time-intensive and requires too much personal investment.

Much effort has gone into devising better measures for scientific success. There is no lack of proposals, but none has caught on because scientists have no good reason to use them. On the other hand, there are many proposals for how to fix the academic system, normally top-down solutions designed to change individual incentives. In neither case however the feedback loop is being closed.

The priority maps close the loop. Now scientists can set their values for certain properties they find desirable of research project or candidates and then they can use their *individual* metrics for assessment. From what the scientists regard important will then naturally arise an aggregated measure that can be used externally. But note that this measure now will automatically adapt as scientists come to regard certain properties (say, number of publications in certain journals) more or less relevant. These measures, importantly, are also non-universal and naturally counteract streamlining. The system can now learn.

2.4 A Five Step Plan

Science matters for steering the future of humanity because it is the tool to obtain knowledge about the systems that govern our lives. Before scientists can use their knowledge to improve social, economic or politic systems, they have to solve their own problems; the academic system is thus the natural starting point. Roughly, the vision I discussed here can be realized in the following 5 steps:

1. **Individual priority maps for scientists**.
   This is possible with presently existing technology. Scientists record their priorities for interesting research and its potential, for characteristics of good science and scientists. The information that their maps are matched to comes from the Science of Science, Scientometrics and Bibliometrics and other sources of knowledge about knowledge discovery. Matches are visualized and can be manipulated to be inspected by the user.
2. **Close the feedback loop in the academic system**.
   The priority maps for scientists will free time for research and make science more efficient. At the same time they provide information about the system itself that can be fed back into the system. Using this knowledge allows the system to learn and creates a naturally adaptive measure for scientific success. The academic system also serves as a case study that reveals difficulties when realizing this feedback loop.
3. **Invidual priority maps for everybody**.
   With the knowledge and, hopefully, the success story from the academic system one can now generalize priority maps for general purpose by adding social, political and personal values. Individuals generate their priority maps and institutions provide correlations.
4. **Close the feedback loop for social systems**.
   Now we can feed back the knowledge about the system into the system and obtain more knowledge by recording the reaction to this feedback. Imagine how much social, political and economic discourse can be shortcut by this. Imagine the creative and engineering potential that is freed.
5. **Upgrade priority maps to brain extensions**.
   In the long run, we should avoid using the visual cortex as pathway to display matches with priority maps. The potential of cheap information will be fully realized when information about our social systems is directly fed back into our brain and we can truly feel the consequences of certain decisions.

2.5 Summary

I assume most people care about the future of the planet as I do, care as you did
with this essay contest asking how humanity should steer the future. I assume most
people want to solve our ecological, political and economic problems, and that we
just have to make it easier for them to convert caring into action. I assume humans
are intrinsically good and mean well, they just don't always get it right.

I may be naïve and I may be wrong. If in fact most people do not regard it relevant
to get the plastic out of the oceans and to prevent children in the developing world
from dehydration, then lowering the cost of information will not make a difference.
However, in that case who am I to tell others that they have to share my values?

I have addressed here the question how humanity can steer the future. I can't tell
you which course we will take if we enable our social systems to learn, but at least
we will not be drifting any more.

References

1. www.fhi.ox.ac.uk. Accessed 8 Apr 2014
2. www.unglobalpulse.org. Accessed 8 Apr 2014
3. www.futurict.eu. Accessed 8 Apr 2014
4. www.radicalism.milcord.com. Accessed 8 Apr 2014
5. Ballew, C.C., Todorov, A.: PNAS **104**, 17948–17953 (2007)
6. www.npr.org/templates/story/story.php?storyId=90327686. Accessed 8 Apr 2014

Chapter 3
Humanity Is *Much* More than the Sum of Humans

Tommaso Bolognesi

Abstract Consider two roughly spherical and coextensive complex systems: the atmosphere and the upper component of the biosphere—humanity. It is well known that, due to a malicious antipodal butterfly, the possibility to accurately forecast the weather is severely limited. Why should it be easier to predict and steer the future of humanity? Here we present various viewpoints on the issue. On a long time scale, we sketch a software-oriented view at the cosmos in all of its components, from space-time to the biosphere and human societies, borrowing ideas from Wolfram, Chaitin and Tononi; this is also motivated by an attempt to provide some formal foundations to Teilhard de Chardin's cosmological/metaphysical visions. On a shorter scale, we discuss the possibility of using formal models, agent based software systems, and big data from social computing, for simulating humanity in-silico, in order to anticipate problems and test solutions.

3.1 Introduction

How should humanity steer its future? This is not precisely the type of problem that I like to mumble about when I wander across the Tuscan countryside. But the arrival, a few days ago, of my 19 year old nephew Tommy—likely, an early member of this future humanity—has triggered a chain of bitter thoughts on the issue.

My solitary walk and meditation have been longer and more winding than usual, this morning, and now that I am back I have a feeling that (i) humanity is not ready to face its stormy future, and (ii) everybody at home has already had their lunch. All except Tommy, of course.

There he is (Fig. 3.1), still in his pyjamas, lying in his preferred location for communication and control: the couch! He must have just moved there from his bed, and is now hyper-connected to a number of electronic devices (laptop, smartphone, iPod, tablet, noise-suppressing headset) that he is using simultaneously, while lazily grabbing the food that my wife has placed near him. Crumbs and chips are scattered

T. Bolognesi (✉)
CNR-ISTI, 1 via Moruzzi, 56124 Pisa, Italy
e-mail: t.bolognesi@isti.cnr.it

© Springer International Publishing Switzerland 2016
A. Aguirre et al. (eds.), *How Should Humanity Steer the Future?*,
The Frontiers Collection, DOI 10.1007/978-3-319-20717-9_3

Fig. 3.1 A node of the
hyperconnected humanity

all over his chest and the floor. What is the future of humanity? Well, part of the
answer is probably under my eyes, right now, and it does not look too promising.

But Tommy must have perceived my presence (how?), because he takes off one
earphone, and, without turning his head:

– What's that, uncle Tomas?
– Tommy! Good morning! Ehm, good afternoon! Did you sleep well?
   And then I cant resist asking him, without preamble:
– Tommy, how should humanity steer its future?

   I guess I am curious to test his sense of self-irony. He does not look surprised:

– Uhm, are we talking like tomorrow or next millennium?
– Well, a couple of centuries might suffice...
– That's fair. So you want the big picture. It's tough! Let me think. Ok: I don't know.
   But you could always ask that french guy...
– Whom?
– What's his name? The author of the book you gave me last summer, remember?
– Ehm. No... I gave you several books...
– Yeah. The guy who invented the Omega thing...
– Oh, you mean Chaitin?
– No.
– Which book?
– The Human Phenomenon [7].
– Aha! Teilhard!
– Sure. Nice guy. He has a lot to say about the future of humanity.
– Like what?

   Tommy [taking off the other earphone and finally looking at me]—Oh come on,
you know it.
   Me—No no, you tell me. That was so long ago, I forgot almost everything.

3.2 Pierre Teilhard de Chardin (1881–1955)

Tommy—Well, he has a whole picture about the evolution of the universe, which is more or less the story of the emergence of life and consciousness. The *outside of things*—matter, that is—evolves from an initial swarm of disaggregated particles to more and more organised forms—atoms, molecules—following the biological law of complexification, or self-organisation. But this growth proceeds jointly with the growth of what he calls the *inside of things*—psyche, thought—which permeates matter from the very beginning, although in rudimentary form. A stone has a soul... but a very small one [smiles]. So the inside and the outside of things develop hands in hands, pushing and pulling each-other, and increasing their complexity, up to the human species, and beyond.

The appearance of the human phenomenon marks the point at which the fabric of the universe achieves the ability to reflect itself. We look outside and we see the physical universe; but we also look deep inside ourselves, and we find ourselves—consciousness. Not only are we conscious; we are conscious to be conscious, which Champ is not [Champ is my dog]. Anyway, to make a long story short, there are four phases: Prelife, Life, Thought, and Superlife. We are now in Thought, and the next stop for humanity, to answer your question, is Superlife, or the Omega point. But I am not sure that we'll get there in the next two centuries, uncle Tomas!

Me—Wow! I start to remember now. Do you realise that he wrote these things around 1940, long before the discoveries of complex and dynamical systems, attractors, notions of emergence, self-organisation phenomena and all that?

Tommy—True. Even more strikingly prophetic, to me, is his concept of Noosphere, the sphere of human knowledge that wraps around planet Earth. It is a sort of biological entity, like the biosphere, but it emerges, on top of that, from the interaction of human minds, and evolves as humanity develops knowledge and cultures, and self-organises into complex social networks. The web of connections becomes thicker and thicker, like neurons in a brain, and becomes increasingly conscious of itself: a super-mind living a superlife. By the way, did you know that the number of people on Earth is only one order of magnitude less than the number of neurons in our brain? Anyway, I think he introduced this vision even before writing 'The Human Phenomenon', long before the Internet, Google, Wolfram Alpha, Facebook, Twitter, and Wikipedia, which I am checking right now... one moment... bingo! Teilhard de Chardin, Noosphere, Cosmogenesis, 1922! And when did they invent the Internet?

Me—Oh, the Arpanet. That was back in...

Tommy—Lets see [checking again his laptop]. Right! 1969: almost 50 years later! How could Teilhard expect the Noosphere to be implemented when the best communication medium of his times was probably the telephone, and there wasn't the slightest clue of all this [he waves his hand to indicate all the technology on the sofa: yes, the boy has a good sense of self-irony!].

Me—So, to go back to my initial question on humanity steering its future, Teilhard's answer might be: keep developing the sphere of human knowledge, the Noosphere, until it becomes a conscious entity— Omega. And then, let Omega steer its own Superlife. It's sad! We are the ants that build the anthill, but the anthill is more than the sum of its ants: it performs complex functions, it lives a life of its own that transcends that of individual ants. How am I, humble ant, supposed to know, share, affect what happens at the upper level? As an ant, I find the picture quite sad!

Tommy—Wait. Teilhard's idea is much more optimistic than that. I think he'd buy the ant-anthill picture, but would place it at a rather early stage in the process of cosmic evolution and emergence of consciousness. You are much more complex than an ant, as a piece of matter, so you also have much more consciousness, since consciousness is a specific effect of complexity. And the difference between you and an ant is not only quantitative, but also qualitative. Let me explain.

Teilhard conceives *two energies* that jointly fuel the complexification process. At any given stage of universal evolution you can see things, simple or complicated, that interact and aggregate under the effect of the *tangential energy*: this is where traditional physics has a lot to say, since we are dealing with the outside of things— their material face. On the contrary, the *radial energy* is in charge of pushing upwards, toward increased complexity, and deals with the inside of things, and their emergent consciousness. In unicellular organisms the focus of organisation F1 (this is his terminology), driven by tangential energy, generates and controls the associated, rudimentary focus of consciousness F2. But, as evolution proceeds, F2 manifests an increasing tendency to mastery and autonomy, and begins to control, with creativity, the dynamics of F1. With the appearance of the human phenomenon, this process reaches a critical point of reversal, where the radial force takes power, so to speak, over matter.

Me—And why should this predominance of F2 over F1 give an advantage to a human over an ant?

Tommy—After the first 'hominisation', leading to *individual* human conscious- ness, a second *collective* hominisation phase develops above us, involving the whole species. With collective hominisation, F2 becomes so strong as to allow humanity to transcend spacetime and join Omega! Consciousness would become coextensive with the universe, whose final state would be pure thought!

Me—Pure thought... and pure philosophy, it seems to me.

Tommy—Well, that's admittedly the most speculative part of the book. In any case, I like the picture of a double-sided final status of the universe: externally it might well be a depressing thermodynamic death, but internally it will be pure thought. And thought is live! Ehm, superlive, to be precise.

Me [Increasingly surprised by Tommy, and increasingly hungry too] Fabulous! Anyway. I am happy that you liked the book! Lets have a serious lunch now!

Tommy—Oh, I did not say that. Strong inspiration, great originality, prophetic visions, but total lack of formalisation, as you pointed out. I do not see how can anyone claim (as he does himself) that the book is a scientific study. Several crucial passages are in desperate need for some mathematical formulation. I remember one in particular... here it is: "*That there are dependencies between the energetics of the inside and the outside of things is incontrovertible. But they can probably be expressed by a complex symbolism described in terms of a different order*". Doesn't it sound like meta-mathematics?

Me—It does! He hints at some mathematical formulation, without actually providing it. Maybe *you* could give a try, Tommy; if I well remember, you already had a course on Thermodynamics, and...

Tommy—I would not even try. Teilhard may have been prophetic in anticipating things that had not yet been invented, but he has missed a fundamental ingredient for his theory that was already available at his times.

Female voice [coming from a videoconference window that has just popped up in Tommy's tablet]—I think I know what you have in mind, Tommy: software!

Tommy [waiving to her]—Alice, good to see you! Good morning! I hope we did not wake you up too early. May I introduce to you my uncle Tomas?

Alice [waiving to me from the small screen]—Nice to meet you. But I did not mean to interrupt you guys. I'll follow your conversation as I take my breakfast here. Please go ahead.

Tommy—Sure! Software. Computation. That's what I meant, and what Teilhard missed. The Turing machine was introduced in 1936, and the notion of algorithm is much older: Teilhard had plenty of time to figure out the key role of computation for understanding the emergence of complexity, the human phenomenon, and for speculating on the future of humanity.

Me [wondering whether the tele-presence of the young female is responsible for the boldness of Tommy's last claim]—Wait! Are you not running too fast? How can you relate the concept of software to the future of humanity?

Tommy—I agree it is not obvious, but... we could start by consulting the Omega guy.

Me—You mean Teilhard de Chardin again?

Tommy—No, I mean Chaitin. Remember the other book that you gave me last summer, Proving Darwin [6]? He has this idea of life as evolving software that I find both simple and powerful. I suppose you've read the book yourself?

Me [lying]—Of course, but that was many years ago. Illuminate me: I forgot almost everything.

Tommy [faking a serious face]—Yeah yeah, except that the book was published in 2012. Anyway...

3.3 Wolfram, Chaitin, Tononi

Tommy—Chaitin's point is that, while we discovered software less than a century ago, Nature invented it a few billion years earlier, when life began to appear on earth. And DNA is the programming language: every single cell in our body contains all the code for building and running a copy of the whole organism. Our DNA code includes stratified traces of our evolution, routines that were operational when we were monkeys, amphibians, or fish! Just like a complex, artificial software system—say an operating system—which evolves by accumulating small increments and patches without ever going through a complete rewriting. The flexibility and plasticity of the biosphere reflect the same properties of software. Spectacular.

Me [definitely hungry, at this point]—Yes, it is. Wow, I am happy that you like the books I give you!

Tommy—I did not say that. What I don't like in Chaitin's view is the line that he draws between rigid, closed, mechanical physics on one hand, and plastic, open, creative biology on the other. He even separates them in terms of the type of math they need: continuous math and differential equations for physics, discrete combinatorial math and algorithms for biology. Now, I certainly agree that biology is creative, to the point that we cannot predict the paths of darwinian evolution in the same way as we predict, in physics, the trajectory of a planet. And I agree that software can be creative more than differential equations. Von Neumann, for example, ended up using discrete maths and computation—cellular automata—for describing self-reproduction, after failing to do so with differential equations. But it is the whole universe to be creative, not just the biosphere!

In his book Chaitin mentions Wolfram and his New Kind of Science [13]. Well, one of the messages from that book is that the emergent properties of the computations of simple programs might explain the complexity and creativity of the physical universe *at all levels*. Spacetime, before anything else, must be creative! And discrete! And algorithmic! Spacetime as a *causal set* [4]—an algorithmic causal set! [2, 3].

Me—But if the universe is creative, and based on software from the bottom to the top, there is no way to predict trajectories, nor the next steps of the human species [and in saying this, I realise that my brain needs some sugar]. By the way, Tommy, lets have some pasta. I did not remember giving you Wolfram's book. When was that?

Tommy—You didn't.

Alice [must have just finished her rich breakfast]—Tommy, I fully agree with you. And let me add that there's another aspect of Chaitin's book in which he seems to underestimate the power of software. He writes that his meta-biology cannot currently address the phenomena of thought and consciousness. These happen to be exactly the pillars of the universal architecture in Teilhard.

Tommy—Interesting. Do we have a proposal?

Alice [Laughing Out Loud]—Do *we* have a proposal? Of course we do. Uhm for what?

Tommy [laughing back]—I mean, for a theory of a digital, computational universe that places software not only at the roots of physics, as Wolfram suggests, and biology, as Chaitin says, but also at the roots of thought and consciousness. Wouldn't this be a nice way to provide a uniform and formal foundation to Teilhard's views?

Alice—Sure! So we need to place thought and consciousness under a software-oriented perspective. Now, if we are ready to accept that sub-atomic particles compute, and bacteria compute, we certainly have no difficulty in admitting that the brain computes. I believe it was Hilary Putnam who said that first, in the 1960's; and that consciousness is computation, regardless of whether it runs in a brain or in a machine. But for placing consciousness in a grand, evolutionary picture, as described by Teilhard, we must find a way to actually *measure* it. And for doing this, I would directly jump to the work of Tononi [1].

Tononi measures the amount of consciousness as the quantity of *integrated information* produced by a complex system. He is a psychiatrist, and his theory supports experimental observations on the human brain and its neural processes, but it is general enough to be applicable, at least in principle, to any kind of system, natural or artificial: neurons in a brain, ants in an anthill, chips on a silicon wafer, computers in a network, humans in a society.

So, imagine a discrete dynamical system S whose global state is characterised by the values of a finite number of discrete state variables, and whose behavior is defined by a state transition law. The amount of *integrated information* produced by S—its degree of consciousness—is the amount of information generated by the system as a whole, beyond the information that its individual parts generate in isolation, by themselves.

Tommy—So, when Aristotle says that the whole is greater than the sum of its parts, we can add that their difference is measured by integrated information?

Alice—I think so. Now, for measuring the information that the system as a whole generates by the very fact of entering some global state X, Tononi considers the reduction of uncertainty produced by that event.

[A new window pops in on Tommy's tablet, where Alice starts typing formulas.] This is defined as the *relative entropy* $H(A\|P)$ between two probabilistic distributions characterizing the state occupied by S *before* entering X: (i) the a priori distribution P (for Potential) of all possible global states, considered as equiprobable (in the absence of any information), and (ii) the a posteriori distribution A (for Actual) that we can infer by knowing X and the state transition law. Think of relative entropy H as a difference between distributions: if the two distributions are identical H is zero; the more they differ, the higher H becomes. This is called *global effective information*.

However, for obtaining *integrated information*—the information that derives from the interactions of system parts, beyond that produced by the parts individually—we must take the difference between A, the *global* actual distribution, and the combination—the product—of all the *local* actual distributions A_i of all individual components, as if they did not cooperate with one another. This is expressed, again, by relative entropy: $H(A\|\prod_{k\in\text{MIP}} A_k)$.

Tommy—MIP?

Alice—Ok, I cheated a bit. You cannot consider an arbitrary partition of the system, but the one that minimizes the difference between the two distributions, to make sure that you have caught the exact delta in information content: MIP is the Minimum Information Partition...

3.4 A Foggy Digital Future

Me [picking the last lonely pop-corn left in the bowl]—Thank you Alice. But I am afraid I can't follow the details now. Let's try to summarise where we are.

Tommy—May I do that? With the help of Teilhard de Chardin, Wolfram, Chaitin, Tononi, and surely many others, for example Seth Lloyd [11], we are trying to interpret the whole universe as a complex *software* system, in the following sense. (1) At the bottom, at the Big Bang, there is a simple, one-line program, capable of setting up a lively spacetime. Spacetime is computed, and itself computes. (2) This spacetime substratum evolves, computationally, and originates more elaborate forms (we obviously want a background-independent theory!), implementing life as we know it—life as evolving software. (3) Then we get humans and their complex, conscious brains. Brains compute. They start networking, and form societies.

And it seems that Integrated Information Theory might be the right technique, at least in principle, for measuring and monitoring the degree of consciousness of each component and composition, at each stage. And yes, Galileo was wrong: the ultimate language of Nature is not mathematics but software.

Me—Wait! Maths has still a lot to say, even about software! Anyway, I see at least two delicate points in this picture. The first is about implementation. Talking about software, we need a program *and* some manipulated data structure, right? What kind of data structure do we have in mind? I would expect it to represent the ultimate fabric of the universe, its 'stuff', as Teilhard would call it—one that the program manipulates and make evolve.

Tommy—That's a somewhat rigid concept of software, uncle Tomas. Ever heard of code that operates on itself? This is commonly done in logic programming. I'd blur the distinction between program and data, and identify one with the other: the program *is* the stuff of the universe, and, at the same time, its engine. Furthermore, a tiny program that sees and modifies itself, at the root of the cosmos, is an elegant way to implement Teilhard's ideas about reflection and the inside of things. Your second point?

Me—No no, I am not convinced. The self-modifying code may be an elegant idea, but if you equate the program with the data structure, thus to the physical universe, you end up with a piece of code as big and complex as the universe itself—a huge program that nobody would ever be able to understand, or even read! The laws of the universe must be understandable! (Who said that?)

Tommy—Albert, but we can't be sure. Anyway, I see your point, and I would rule out this gigantic monolithic piece of code too. This is perhaps how some badly designed, man-made software systems grow. But I imagine a different scenario, inspired by the biosphere: a small piece of code that replicates, like cellular automata can do, and mutates, much like in genetic programming. We'd have populations of simple computing entities, first stateless and then stateful, that start to interact with each other, building up complexity in layers, by emergence, in line with Teilhard de Chardin, and even Darwin! It is still a software-centric universe, but distributed: not the nightmare that you mention. The good news is that, being software, we can run it in a simulation. The bad news: we cannot hope to reproduce but a tiny incipit of our history.

Alice—But Tommy, you might remember that argument by Wolfram: if a simulated universe, starting from a totally abstract piece of code, managed to faithfully predict just two or three known physical phenomena, or constants, then you would have very strong reasons to believe that you have caught all the rest of physics. So there is still some hope!

Fig. 3.2 Foggy universe from two lines of code

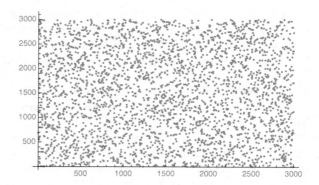

Me—Maybe some hope to find that software; certainly not to use it for predicting the future of humanity. Anyway, until these two or three predictions are not there, I guess this idea of a natural universe emerging from *two lines of code*, nicely minimalistic as it can be, should be put on the same shelf as Teilhard's book—the shelf of *foggy* cosmologies, if you know what I mean.

Champ [coming in from the garden, with muddy legs]—woof, woof!

Tommy [quickly disconnecting from all his hardware, and jumping down the couch]—Don't worry uncle Tomas, it is my turn. Alice! I'll catch you later!

Tommy walks out with the dog. I say bye to Alice too. Her window pops out, revealing a smaller window with just *two lines of code*:

```
step[{array_, pos_}] := {Insert[array, Length[array] + 1, pos], array[[pos]]}
ListPlot[Nest[step, {{1, 2}, 2}, 3000][[1]]]
```

I can't resist hitting Control-Enter. The code runs. A foggy universe indeed (Fig. 3.2).

3.5 Humanity In Silico

Consider two roughly spherical and coextensive complex systems: the atmosphere and the upper component of the biosphere—humanity. It is well known that the possibility to accurately forecast the weather is severely limited. Why should it be easier to predict and steer the future of humanity? In this closing section we present both pessimistic and optimistic arguments about the possibility to effectively predict and drive our future on reasonable time scales, shorter than those implied by the above conversation between Tomas, Tommy and Alice.

Humanity is a complex network of networks—groups, societies—of individuals. The network is multi-layered, and the individuals are diversified and complex stateful systems themselves. Not only they interact with one another but, crucially, they compete for accessing global resources whose scarcity has boosted the evolution of the species but may soon become a threat for its survival.

Consider now a much simpler system of interacting entities: elementary cellular automata (CA) [13]. They are single layered. Individuals (binary cells) are all equal and follow the same elementary behavioural rule. Interactions are short-range. There's no environment. And yet, for the most interesting CA of the family (and our universe is most interesting!) the long term evolution of the system cannot be predicted other than by executing it step by step. It would seem that the pure ability to *interact*, even in a limited way, is enough for triggering complex and unpredictable behavior.

How about controllability? By changing the behavior of a single cell in a CA, an avalanche of modifications causally spreads across the spacetime diagram. A binary cell cannot decide to flip itself, or change the boolean function that defines its behavior; but humans can. Experiments with CA show that, in the most interesting cases (see Fig. 3.3-left, and demonstration [14]), the pattern obtained by tracing the consequences of a flipped cell appears itself as the spacetime diagram of an unpredictable, irreducible computation, even though patches of regular, predictable behavior may appear from time to time (Fig. 3.3-right).

Are there more effective mechanisms for steering the spacetime diagram of humanity? For doing this, we set up local administrations, governments, international bodies that promulgate and enforce laws and regulations. But, viewing these human groups/societies themselves as upper level *individual entities* that interact with one another, we seem to be still dealing with a complex, unmanageable system. One may argue that the situation at this upper level is different because, say, the text of a law represents a form of communication that spans widely across space and time. However, experiments with non-elementary CA show that widening the spatial and temporal context that defines the actions of individual cells does not substantially modify the spectrum of possible overall behaviors.

Although the lessons from CA—perhaps the most basic model of populations of interacting entities—are not encouraging, more sophisticated models have been developed, often inspired by CA, that justify some optimism.

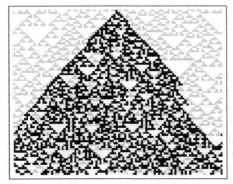

Fig. 3.3 Spreading effect of a perturbation in elementary automata 22 and 54

In Kaufmann's *boolean networks* [9], a refined version of CA, the rigid structure of the cell grid is replaced by a graph structure, and the agents' behaviors are not uniform. Under rather lax conditions, these networks exhibit ordered emergent behaviors—a form of 'order for free' that occurs when the dynamics is trapped within an attractor covering a tiny fraction of the global state space. Kauffman's networks have been successfully used for modelling the emergence of order (e.g. metabolic functions) in molecule soups. Humans are social *atoms* that aggregate in social *molecules*; the 'chemical' reactions in which we engage are monitored and recorded day by day, by social networks and a variety of other devices. Beside the current commercial use, 'big data' could be used for predicting trends, anticipating problems and testing solutions at the global scale.

In fact, advances in *agent based modelling* and *agent based social simulation* (see [5, 10] for reviews), combined with the diffusion of social networks and an increasing academic interest in their analysis, have led to more and more accurate models of (aspects of) the human network and its environment. Among the pioneers of this research area, Schelling, inspired by CA, has developed the first agent-based simulation for studying social phenomena—in particular, segregation patterns - over 40 years ago [12].

Schelling's ideas were further developed by Joshua M. Epstein and Robert Axtell [8], who developed the *Sugarscape* model. The model originally consisted of agents living in a 51×51 grid whose cells contain different amounts of sugar or spice, and of rules for sophisticated agent behaviours, including interactions with other agents and with the environment. Several implementations of the Sugarscape model are currently available as open source software (see http://en.wikipedia.org/wiki/Sugarscape).

Several other packages are available for the simulation of populations up to millions of agents, and for the study of their emergent properties. These include *Swarm* (http://savannah.nongnu.org/projects/swarm), *Casos* (http://www.casos.cs.cmu.edu/index.php), *Repast* (http://repast.sourceforge.net/), *GAMA* (https://code.google.com/p/gama-platform/).

References

1. Balduzzi, D., Tononi, G.: Integrated information in discrete dynamical systems: Motivation and theoretical framework. PLOS Comput. Biol. **4**(6), e1000091 (2008)
2. Bolognesi, T.: Algorithmic causets. In *Space, Time, Matter - current issues in quantum mechanics and beyond—Proceedings of DICE 2010*. IOP, J. Phys.—Conf. Ser. (2011)
3. Bolognesi, T.: Causal sets from simple models of computation. Int. J. Unconv. Comput. (IJUC), 7, 2011
4. Bombelli, L., Lee, J., Meyer, D., Sorkin, R.D.: Space-time as a causal set. Phys. Rev. Lett. **59**(5), 521–524 (1987)
5. Bonabeau, E.: Agent-based modeling: Methods and techniques for simulating human systems. Proc. Nat. Acad. Sci. **99**(3), 7280–7287 (2002)
6. Chaitin, G.: Proving Darwin—Making Biology Mathematical. Pantheon Books, New York (2012)

7. de Chardin, T.: *Le Phénomène Humain*. Ed. du Seuil, Paris (1955)
8. Epstein, J.M., Axtell, R.L.: Growing Artificial Societies: Social Science from the Bottom Up. MIT Press, Cambridge (1996)
9. Kauffman, S.A.: At Home in the Universe: The Search for Laws of Self-organization and Complexity. Oxford University Press, Oxford paperbacks (1995)
10. Li, X., Mao, W., Zeng, D., Wang, F.-Y.: Agent-based social simulation and modeling in social computing. In: Yang, C.C., et al. (eds.) Intelligence and Security Informatics. Lecture Notes in Computer Science, vol. 5075, pp. 401–412. Springer, Berlin Heidelberg (2008)
11. Lloyd, S.: Universe as quantum computer. Complexity **3**(1), 32–35 (1997)
12. Schelling, T.: Dynamic models of segregation. J. Math. Sociol. **1**, 143–186 (1971)
13. Wolfram, S.: A New Kind of Science. Wolfram Media, Inc. (2002)
14. Zenil, H., Villarreal-Zapata, E.: Sensitivity to peturbation in elementary cellular automata. The Wolfram Demonstrations Project. http://demonstrations.wolfram.com/SensitivityToPeturbationInElementaryCellularAutomata/

Chapter 4
Crucial Phenomena

Daniel Dewey

Abstract I give a case that, as a public good, societies and their governments should support and invest in scientific research on *crucial phenomena,* empirical features of the world that figure strongly in how humanity's choices influence the size of its future. In particular, I give reasons for thinking that (1) humanity's vulnerability or robustness to accidents arising from biological engineering, and (2) the future rates of improvement of artificial intelligence and its susceptibility to misuse, are phenomena that call strongly for our systematic attention.

4.1 Introduction

In his 1986 talk *You and Your Research*, Richard Hamming recounts a story about his time at Bell Labs:

> Over on the other side of the dining hall was a chemistry table. I had worked with one of the fellows, Dave McCall; furthermore he was courting our secretary at the time. I went over and said, "Do you mind if I join you?" They can't say no, so I started eating with them for a while. And I started asking, "What are the important problems of your field?" And after a week or so, "What important problems are you working on?" And after some more time I came in one day and said, "If what you are doing is not important, and if you don't think it is going to lead to something important, why are you at Bell Labs working on it?" I wasn't welcomed after that; I had to find somebody else to eat with! That was in the spring [9].

The individual researcher will often have practical answers for Hamming's final question: funding may not be available for important problems in one's field, or one might not have the particular skills and resources that would be required to tackle those problems. However, it is not so easy to escape from Hamming's questions

Supported by the Alexander Tamas Research Fellowship on Machine Superintelligence and the Future of AI.

D. Dewey (✉)
Oxford Martin Programme on the Impacts of Future Technology,
Future of Humanity Institute, Oxford, UK
e-mail: daniel.dewey@philosophy.ox.ac.uk

© Springer International Publishing Switzerland 2016
A. Aguirre et al. (eds.), *How Should Humanity Steer the Future?,*
The Frontiers Collection, DOI 10.1007/978-3-319-20717-9_4

when they are asked of the entire scientific community, or of society as a whole: what are the important problems of our time, and what problems are we working on? If resources are misallocated or the required skills are not available, then we have nobody to blame for this but ourselves.

This essay puts forth the idea that one of the most important tasks facing us today is the scientific investigation of certain *crucial phenomena*, empirical features of the world that figure strongly in how humanity's choices influence the size of its future. Technical, publicly understandable knowledge of these phenomena is important in two senses: it has great value in terms of consequences, and it seems to be reasonably achievable from our current position (in Hamming's terminology, we "have an attack" on these phenomena). In particular, I will give reasons for thinking that (1) humanity's vulnerability or robustness to accidents arising from biological engineering, and (2) the future rates of improvement of artificial intelligence and its susceptibility to misuse, are phenomena that call strongly for our systematic attention.

I begin with a series of arguments culminating in an moral rule of thumb that we ought to maximize the chance that humanity's future is "Large" instead of "Small"; then, I show the relevance of crucial phenomena to this endeavor. Finally, I put forward the claim that since in many cases we do not know enough about crucial phenomena to make good decisions, we ought to be working towards scientific knowledge of crucial phenomena, and that we should focus on the most time-sensitive ones.

This essay is built of significant insights from several people, relying particularly heavily on the ideas of Nick Bostrom and Nick Beckstead. My incremental contribution is to compile these ideas into a form that makes the value and urgency of certain kinds of scientific knowledge clear, and to argue that the acquisition of this knowledge is one of the best available policies for humanity today.

4.2 Aim for a Large Future

Significant credence must be given to the idea that many times more potential value lies in humanity's long-term future—let's say the year 2100 and onward—than lies in its short-term future, between now and 2100. One way of supporting this proposition is to combine the following three plausible premises, two moral and one empirical:

1. Value is likely to be aggregative: more of a good thing is more valuable, and returns do not diminish quickly. "Good things" could be happy experiences, virtuous people, beautiful works of art, etc.
2. Intrinsic value is likely to be time-insensitive: whether a thing exists in the year 571 BC, 2014 AD, or 20014 AD does not affect its intrinsic value.
3. Humanity's long-term future has the potential to contain vastly more good things than its near-term future.

The third premise is the empirical one, following from the idea that humanity's long-term future could plausibly be *many times larger* than its near-term future. It could be many times larger in duration, since the universe is expected to continue in a usable state for at least on the order of billions of years. It could also be many times larger in 'breadth', roughly the number of 'things' (people, artifacts, etc.) humanity influences at any given time, as increasing technological ability grants us increasing access to the resources of Earth, our solar system, our galaxy, and so on outwards to larger stages.

It follows, then, that if we can meaningfully affect humanity's long-term future, then it is immensely important that we do so. It is plausibly much, much more important to influence humanity's future from 2100 onward, than it is to influence the mere 86 years we have remaining between now and 2100. I will not try to argue the point conclusively here, since there are many subtleties and others have done so much better; I refer the curious reader especially to Beckstead [2].

How could we meaningfully affect humanity's long-term future? Nick Bostrom approaches the problem by pointing to the category of *existential risks*, risks that threaten "the premature extinction of Earth-originating life or the permanent and drastic reduction of its potential for desirable future development" [4]. By definition, any risk of the loss of a significant part of humanity's long-term future is covered by Bostrom's concept of existential risk. Given this definition of existential risk, Bostrom argues that it may be useful to adopt a moral rule of thumb which he calls "maxipok": "Maximize the probability of an "OK outcome," where an OK outcome is any outcome that avoids existential catastrophe."

Technically, the second clause of Bostrom's definition of existential risk renders the first redundant; extinction is an example of an event that could permanently and drastically reduce humanity's potential for desirable future development. In another writing,[1] Bostrom sheds some light on this choice of emphasis:

> The notion [of existential risk] is more useful to the extent that likely scenarios fall relatively sharply into two distinct categories—very good ones and very bad ones. To the extent that there is a wide range of scenarios that are roughly equally plausible and that vary continuously in the degree to which the trajectory is good, the existential risk concept will be a less useful tool for thinking about our choices. One would then have to resort to a more complicated calculation. However, extinction is quite dichotomous, and there is also a thought that many sufficiently good future civilizations would over time asymptote to the optimal track.

In other words, the concept of existential risk is most useful if futures can be roughly sorted into two categories, *Prosperous* and *Disastrous*; human extinction, being dichotomous, supports this sorting hypothesis.

In the definition of existential risk, what is this "desirable future development" that could be drastically reduced? Though this depends on open questions in moral philosophy, it seems to me that we can approximate the desired quantity by the *size* of humanity's future, as defined in Sect. 4.2—humanity's duration in time, times its 'breadth' in terms of the matter and energy (and other resources) controlled by

---

[1] This quote appeared as a comment on Beckstead [1].

humans.[2] In this case, one could simply consider the *size* of the future to get a fairly good guide for how desirable it is. Does this "size view" fit with the existential risk picture? Let us examine the four categories of Disastrous (existentially catastrophic) futures that Bostrom lists: human extinction, permanent stagnation, flawed realization, and subsequent ruination.

Extinction: Presumably since extinction eliminates anything that could be considered "future development" at all, Bostrom does not further explain extinction's impact on existential risk.

Permanent stagnation: The value lost through permanent stagnation is exemplified by these futures' inability to "produce astronomical numbers of extremely long and valuable lives"—to create a 'broad' future, in my term from above.

Flawed realization: Flawed realization comes in two varieties, "unconsummated" and "ephemeral". Unconsummated realization covers disasters in which something critical about value has been lost on the way to technological maturity; for example, humans may have replaced themselves with artificially intelligent machines, but accidentally failed to make these machines so that they could have phenomenal experience, resulting in a hugely broad, long future with "no morally relevant beings there to enjoy the wealth". In an ephemeral realization, humanity crashes down to extinction or permanent stagnation shortly after reaching technological maturity.

Subsequent ruination: Subsequent ruination refers to futures in which humans reach technological maturity with all critical aspects of value intact, but through failure of luck or wisdom do not realize much of the potential desirable future development. Bostrom emphasizes that this situation seems less likely, given the considerable resources at humanity's disposal, and that we are not in much of a position to help these future people in any case; they seem to have been given all of the advantages we could hope to give them.

In all but one of these cases (unconsummated realization), the badness of existential risk, the "future development" that is lost, can be accounted for in terms of the size of humanity's future, its breadth and duration, without considering other qualities. Given this observation, one could propose a new way of characterizing Bostrom's "very good" and "very bad" categories of futures. The hypothesis would be that plausible futures of humanity naturally split into Prosperous "Large" futures, where humanity's future has a long duration and is broad in terms of resources controlled, and Disastrous "Small" futures, in which humanity is either short-lived or "narrow", having relatively little control over resources.

This in turn suggests a new rule of thumb, in parallel to Bostrom's maxipok rule:

Aim for Large: Maximize the probability of humanity's future being Large instead of Small.

---

[2]This is not to say that size is intrinsically valuable. The assumption is that future humans will figure out how to do something good, given enough size.

Relative to maxipok, aim-for-large's advantage is that it cashes out normative language—"potential for desirable future development"—in terms of the relatively concrete duration and breadth of (i.e. resources controlled by) humanity's future.[3]

This is certainly not to say that all Large futures are good, or are better than all Small futures, just that a future's size is an unusually useful piece of information about how good a future is. Choosing to focus only on whether humanity's future is Large or Small loses some nuance; the example of an unconsummated realization, in which humanity's future is long and broad but lacks value, makes this clear. However, for practical purposes, there is much to be gained by cashing out normative language in concrete terms.

4.3 Crucial Phenomena

The aim-for-large rule leaves us with a new question: how *can* humanity act to maximize the probability of having a Large future? We cannot wish a chosen future into existence; instead, our choices interact with features of the world, and the fundamental and emergent laws that govern those features determine how our choices affect humanity's future's duration and breadth. Our ability to choose effectively depends on our knowledge of these empirical phenomena.

Crucial phenomena are empirical phenomena that play a key role in determining how humanity's actions influence whether its future is Large or Small. By *empirical phenomena*, I mean relationships that hold between sets of real-world conditions. "The moon waxes and wanes in such-and-such a pattern" is an empirical phenomenon. Physics-based phenomena such as the phase transitions of water, emergent phenomena such as the relationships between predator and prey populations, and mathematical "phenomena" that become realized in the world, such as the difficulty of factoring a large composite number found encoded on a hard-drive, are also empirical phenomena. This broad usage is meant to capture all kinds of patterns that are found in features and behaviours of the world.

Some phenomena are much more relevant than others in determining whether humanity's future is Large or Small. For example, while different laws of plate tectonics could result in dramatic differences in future arrangements of planetary oceans and landmasses, it seems unlikely that these differences would result, *ceteris paribus*, in significant differences in humanity's duration or breadth. On the other hand, humanity's future size could be dramatically impacted by the cosmological rate of expansion, which determines how much matter is ultimately reachable by humans. Different cosmologies have radically differently-sized futures of humanity [7].

---

[3]It would be misleading to say that aim-for-large does not contain any non-concrete or normative language; the definition of "humanity" plays a key role in defining the depth and breadth of humanity's future, and the definition of "humanity" is surely normatively loaded and subject to debate.

It may be instructive to imagine that humanity's future is determined by a game whose players are Humanity and Nature. Each player has a number of parameters that they are allowed to set; Humanity's parameters correspond to its choices, and Nature's parameters correspond to empirical phenomena. There are some of Nature's choices that will affect the outcome of the game greatly, and some that will do so in such a way that Humanity would benefit greatly from being given a peek at Nature's move; given the knowledge of how Nature sets its phenomena, Humanity could act to maximize the value of their play.

One relatively natural categorization of crucial phenomena I have found splits the full set into four subsets, induced by two binary qualities: each phenomenon affects either primarily *duration* or *breadth*, and does so in a way that is either *limiting* or *transformative*. These distinctions can be best understood by enumerating the four categories:

Duration-limiting phenomena: Duration-limiting phenomena are crucial by virtue of their potential to limit the possible duration of humanity's future; they set bounds on how long or short our future could be, often in ways that our choices cannot affect. For example, vacuum collapse could act as a duration-limiting phenomenon, as could physical factors such as the decay time of protons, or cosmological factors such as the time until a big crunch or similar universe-ending event.

Breadth-limiting phenomena: Breadth-limiting phenomena are crucial by virtue of their potential to limit the possible breadth of humanity's future. For example, phenomena that determine the material cost of taking control of additional solar systems (the density of interstellar dust, failure rates of relevant technologies, etc.) affect the potential breadth of humanity's future, as do some cosmological factors such as the rate of expansion. Laws of physical computing efficiency could also act as limiting factors on breadth, if computation is particularly relevant to the kinds of things we'd want to create in our future.

In most cases, crucial *limiting* phenomena—whether duration-limiting or breadth-limiting—don't interact *directly* with humanity's choices so much as they place boundaries on the stage on which human choices will be played out. We cannot choose in ways that change these basic limitations, but we can react to limitations to make sure the future is Large instead of Small. For example, the optimal trade-off between speed and caution of development could depend on how long we think we have and how widely we should expect to spread. If one overestimates the amount of time left, then some investments may be left to gather interest for too long, resulting in a suboptimal payout of value.

Duration-transformative phenomena: Duration-transformative phenomena are crucial because they directly determine some mapping between choices and durations in a dramatic way. For example, the phenomena harnessed by technologies that carry extinction risks are duration-transformative; the details of physics phenomena determine whether the action of activating a particle accelerator will yield useful scientific knowledge, or alternatively create a strangelet that converts the

Earth into a lifeless lump of strange matter. Preventable natural extinction risks also fall into this category.

Breadth-transformative phenomena: These are phenomena that are crucial because they directly determine some mapping between choices and breadths in a dramatic way. For example, the potential of von Neumann probes to waste large chunks of the cosmic resource pool and the effects of anti-space-colonization memes both create mappings between some of our choices and humanity's future breadth. Whatever phenomena ultimately explain Fermi's Paradox may turn out to be breadth-transformative phenomena.

4.4 Steering the Future

I have derived the term "crucial phenomena" from Bostrom's *crucial considerations*, "idea[s] or argument[s] that might plausibly reveal the need for not just some minor course adjustment in our practical endeavours but a major change of direction or priority" [3]. Since crucial phenomena are so important to our future, knowledge of the existence of a crucial phenomenon, of the laws that govern it, or of the ways that it interacts with our choices, will sometimes be crucial considerations.

Crucial phenomena relate to the aim-for-large rule in a simple way: whenever we face a choice, we ought to use whatever knowledge we have of crucial phenomena in order to choose the option that maximizes the chance of a Large future. By the time a given choice is presented to us, we should do our best to have the required knowledge of whatever crucial phenomena will be relevant to that choice well in hand.

Given that some piece of knowledge arrives in time to inform a particular choice, what properties would it be best for knowledge of crucial phenomena to have? An obvious first step would be that it should be *reliable*. Another desirable property is that the relevant knowledge should be *permissible as grounds for decisions that affect the common good*. While it may be acceptable to act on hunches or private evidence when making decisions on one's own behalf, it would be best if knowledge that guides significant decisions about humanity's shared future could be based on publicly verifiable evidence. This criterion is especially important given that governments will likely play significant roles in decisions that would benefit from knowledge of crucial phenomena; it would thus be desirable that our knowledge of crucial phenomena be available to them in a form that they can use legitimately.

Fortunately, we have societal means to secure reliable knowledge that is publicly verifiable and usable in common-good decision-making: the scientific community. The formal, professional, and social structures that make up the modern practice of science have been extremely effective in advancing our knowledge of many phenomena and in allowing us to harness and control those phenomena to improve our lives. Science can achieve the high reliability we need, and can be publicly examined and sanctioned as evidence to be used in making decisions that affect humanity's future in significant ways.

Thus, we come finally to a recommendation: as a public good, societies and their governments should support and invest in scientific research on crucial phenomena, prioritized according to the estimated size of their impact and the nearness of the relevant decisions we will need to make. I have taken such trouble to give my reasons for supporting this position because on the surface, it may sound familiar: unsurprisingly, researchers often declare that "funding for further research is needed"! To the extent that you have found my arguments convincing, however, this psychological explanation should not debunk the real need for this *particular* kind of "further research". If we are to make reliable, effective policy decisions in the future, then we must make a policy decision now to invest in our understanding of crucial phenomena.

To be as concrete as possible, I will describe two crucial phenomena that are relevant to decisions that are happening either soon or in the present time. These phenomena are (1) humanity's vulnerability or robustness to accidents arising from biological engineering, which I will call "biological instability", and (2) the future rates of improvement of artificial intelligence and its susceptibility to misuse, which I will call "AI improvement and misuse properties".

Biological instability: Humanity is constituted of and embedded in biological systems, and biological engineering is advancing at a rapid rate. How unstable is humanity, or the ecosystem that we depend on, in the face of novel agents that could be produced by biological engineers? It seems from the historical record that it would be relatively difficult for natural mutation to stumble on an organism that could render humanity extinct[4]; is this because the space of biological organisms contains few of such threats, or is this merely a property of the part of the space that Nature can easily explore?

Factors that determine humanity's robustness against artificial biological system shocks—facts about the difference between the spaces of probable natural and artificial agents, facts about epidemiology of artificial agents, facts about the difficulty or ease of an ecosystem "takeover" by engineered organisms—are duration-transformative crucial phenomena, which could determine whether many actions we take are relatively harmless, or whether they could render humanity's future Small through extinction.

AI improvement and misuse properties: Though we cannot claim knowledge of specific future techniques in the field of artificial intelligence, there are reasons to look at the general "landscape" of artificial intelligence and conclude that large, sudden jumps in AI capabilities and rates of improvement are plausible. First, if several conjunctive factors all must reach a certain level before some capability is achieved, then some factors may reach many times the required level before the final factor reaches the critical level; at that point, the system will fall up an "overhang", suddenly achieving a much more effective, efficient, or economical version of the desired ability [12, 13]. Second, there is a set of cognitive skills, exemplified in human children, that can be used to learn many other cognitive skills directly from human cultural artifacts such as books and websites; we should expect a large jump in capability as AI systems quickly

---

[4]Though see Ćirković et al. [6] for commentary on the relevant concept of "anthropic shadow".

acquire any cultural skills that have not yet been automated. Finally, and perhaps most significantly, it is clear that AI research and development is a cognitive skill like any other, and should be subject to automation; when it is, there are reasons to think that the rate of improvement of AI capabilities would accelerate sharply upward in an "intelligence explosion" [8, 15]. It is plausible that AI could reach levels of capability far beyond any human or group of humans at any number of tasks.

Furthermore, it has been argued that if we attempt to use such "superintelligent" AI, or if an AI system were to improve to a high level while in use, it would be easy to accidentally *misuse* such a system, and that such accidental misuse could lead to results as extreme as the extinction of humanity. This idea has been given a few supporting arguments, including the existence of "convergent instrumental goals" that would cause AIs with many different tasks to take actions that could harm humanity [5, 11] and the difficulty of designing tasks for superintelligent AIs that would result in non-Disastrous outcomes [14].

The phenomena that govern AI improvement rates and its susceptibility to misuse, which I have outlined above, could lead to human extinction, and so they are duration-transformative crucial phenomena. Depending on the true nature of these phenomena, certain kinds of AI research and development in the medium-term future could threaten human extinction, or could be purely beneficial ways of creating helpful new technology.

Of these two, it is plausible that biological instability is the more urgent crucial phenomenon. While AI still appears to be far from the two thresholds that I mention, biological engineering is creating novel, harmful agents today, and escapes from BSL-4 (highest security) labs are shockingly common [10]. Additionally, opportunities to monitor or regulate new biological technologies before they become too widespread for effective control may soon slip through our grasp. On the other hand, threats from AI improvement and misuse have a more deeply puzzling character; even if we understood them well, it is not clear what appealing policy courses we could take to mitigate them. Since AI improvement and misuse may require significantly more work to solve (should it prove to be a true problem), it should also be treated with some urgency.

In this essay, I have sought to explain why societies and their governments should support and invest in scientific research on crucial phenomena. In particular, there are common-good issues where we lack sufficient understanding to take more proactive policy action; in these cases, such as the case of biological instability, engaging in scientific research may be the best policy choice available. Hamming, once again:

> Our society frowns on people who set out to do really good work. You're not supposed to; luck is supposed to descend on you and you do great things by chance. Well, that's a kind of dumb thing to say. I say, why shouldn't you set out to do something significant. You don't have to tell other people, but shouldn't you say to yourself, "Yes, I would like to do something significant."

This quote could be applied not just to individuals, but to generations and societies. In the case of humanity's long-term prospects, our collective humility and our duty to the future are in conflict with with one another. "So much the worse for our collective humility" seems, to me, the only acceptable response.

References

1. Beckstead, N.: A proposed adjustment to the astronomical waste argument. LessWrong post. http://lesswrong.com/lw/hjb/a_proposed_adjustment_to_the_astronomical_waste/ (2013)
2. Beckstead, N.: On the overwhelming importance of shaping the far future. Ph.D. thesis, Rutgers University (2013)
3. Bostrom, N.: Crucial considerations. http://nickbostrom.com/
4. Bostrom, N.: Existential risk prevention as global priority. Glob. Policy 4(1), 15–31 (2013)
5. Bostrom, N.: The superintelligent will: motivation and instrumental rationality in advanced artificial agents. Minds Mach. 22(2), 71–85 (2012)
6. Ćirković, M.M., Sandberg, A., and Bostrom, N.: Anthropic shadow: observation selection effects and human extinction risks. Risk Anal. 30(10), 1495–1506 (2010)
7. Ćirković, M.M.: Cosmological forecast and its practical significance. J. Evol. Technol. 12 (2002)
8. Good, I.J.: Speculations concerning the first ultraintelligent machine. Adv. Comput. 6(31), 88 (1965)
9. Hamming, R.W and Kaiser, J.F.: You and your research. In: Transcription of the Bell Communications Research Colloquium Seminar (1986)
10. Klotz, L.C., and Sylvester, E.J.: The unacceptable risks of a man-made pandemic. Bull. At. Sci. 7 (2012)
11. Omohundro, S.M.: The basic AI drives. In: Frontiers in Artificial Intelligence and Applications, Vol. 171, p. 483 (2008)
12. Shulman, C., Sandberg, A.: Implications of a software-limited singularity. In: Proceedings of the European Conference of Computing and Philosophy. (2010)
13. Yudkowsky, E.: Artificial intelligence as a positive and negative factor in global risk. In: Global Catastrophic Risks, Vol. 1, p. 303 (2008)
14. Yudkowsky, E.: Complex value systems in friendly AI. In: Artificial General Intelligence, pp. 388-393. Springer (2011)
15. Yudkowsky, E.: Intelligence explosion microeconomics. Technical report, 2013-1. Machine Intelligence Research Institute, Berkeley, CA. http://intelligence.org/files/IEM.pdf

Chapter 5
The Leverage and Centrality of Mind

Preston Estep and Alexander Hoekstra

Abstract Humanity faces many critical challenges, many of which grow relent-lessly in seriousness and complexity: declining quantities and quality of freshwater, topsoil, and energy; climate change and increasingly unpredictable weather patterns; environmental and habitat decline; the growing geographical spread and antibiotic resistance of pathogens; increasing burdens of disease and health care expenditures; and so on. Some of the most serious problems remain intractable, irrespective of national wealth and achievement. Even developed nations suffer from stubbornly stable levels of mental illness, poverty, and homelessness, in otherwise increasingly wealthy economies. A known root cause of such broken lives is broken minds. What isn't widely recognized is that all other extremely serious problems are similarly and equally intertwined with the intrinsic incapacities of human minds—minds evolved for a focus on the short term in a slower and simpler time. Yet minds are also simul-taneously the most essential resource worth saving, and the only resource capable of planning and executing initial steps of necessary solutions. There is hope for overcoming all serious challenges currently facing us, and those on the horizon; yet there is only one most-efficient strategy that applies to them all. This strategy focuses not on these individual and disparate challenges—which ultimately are only symptoms—but on fixing and improving minds.

5.1 Background

Humanity faces many serious challenges. Some already appear imposing, yet grow relentlessly in seriousness and complexity. Critical resources are in decline in much of the world. Quantities and qualities of clean air, fresh water, and topsoil are dimin-ishing [1]. Production levels of critical non-renewable resources (such as oil) have peaked in most of the world. Climate change and increasingly unpredictable weather

P. Estep (✉)
Mind First Foundation, 504, Lincoln, MA 01773, USA
e-mail: pwestep@gmail.com

A. Hoekstra
11 Willard Rd., Weston, MA 02493, USA

© Springer International Publishing Switzerland 2016
A. Aguirre et al. (eds.), *How Should Humanity Steer the Future?*,
The Frontiers Collection, DOI 10.1007/978-3-319-20717-9_5

patterns make regular news. Border skirmishes and wars still break out routinely in many areas of the world. Other notable challenges include environmental and habitat decline, the growing geographical spread and antibiotic resistance of pathogens, increasing burdens of disease (especially in growing numbers of elderly) and health care expenditures, a potentially catastrophic asteroid strike, and so on.

5.2 Immature Science

A recent poll by The Pew Research Center shows that most in the U.S. expect science and technology to come to the rescue—a view likely shared by an increasing number of people in other countries [2]. Although those polled have a favorable view of technological progress generally, the poll also indicates that many specific advances are regarded with suspicion or even trepidation. This dichotomy reveals the uneasy historical relationship between a general perceived need for betterment, and the implementation of potentially disruptive specific ideas or technologies. Even the practice of science itself had trouble gaining initial traction, since it historically required that a single individual propose a new idea that challenged prevailing orthodoxy.

Modern discoveries in genetics show us that human populations separated and have lived in essential isolation from each other for at least 50,000 years, and we know that people from all separated branches of the family tree are able to do science [3]. It is very unlikely that separated human populations experienced universal convergent evolution toward scientific ability, and much more likely that humans at that time of divergence were capable of science. Yet the age of modern science is probably less than 500 years old—only about 1 % of the time since populations split. Understanding why science is so unnatural, and took so long tells us much about human nature and our inherent resistance to change. It also helps us chart our best possible course to the future.

Science and engineering are considered inseparably intertwined in the modern world, but this hasn't always been so. Engineering was quite advanced prior to modern science. For several thousand years, humans have been designing and building amazingly complex and sophisticated roads, bridges, aqueducts, buildings and amphitheaters. Consider the Egyptian pyramids—feats of exceptional engineering. They are over 4500 years old, and even far older monuments and artifacts stand as persuasive testimony to the very long history of engineering. Effective tools and weapons were being made well over 1 million years ago. So why is science so young? Let's begin at the official beginning.

Though exact dates are disputed it is a generally held convention that the year 1543 launched the Scientific Revolution [4]. Andreas Vesalius published the first work of scientific physiology and Nicolaus Copernicus published his revolutionary claim that the earth orbited the sun, rather than the other way round. Copernicus withheld publication of his heliocentric theory for many years—until 1543, the year of his death—because he feared the repercussions. Copernicus had very good reason to fear, and even if he'd lived another century he might have chosen the same course. Galileo Galilei's observational evidence from the early 1600s in support of the Copernican theory was dealt with harshly by the Roman Catholic Church, and he spent almost the last decade of his life under house arrest, dying in 1642. Important advances in science and mathematics were made throughout Europe for the remainder of the 17th century, most notably by Sir Isaac Newton, but Newton and other scientists were very guarded about their religious views and were very careful to explain away any possible contradictions their findings might present to accepted religious orthodoxy. In 1697 Thomas Aikenhead was the last person hanged for blasphemy in Britain. The 18th century brought more but still slow and gradual change in the perceptions of science.

Over two centuries after Galileo's death, and a century and a half after Aikenhead's execution, Charles Darwin—like Copernicus three centuries before—feared the repercussions of his revolutionary ideas, and delayed publication for as long as possible. Darwin might have followed Copernicus' example, and waited until death was imminent to publish his theory, but a letter from Alfred Russel Wallace, describing his own formulation of essentially the same theory, compelled Darwin to publish. He did so fretfully, fully aware of the still-restrictive social climate and history of persecution—and even execution—of those who dared contradict official church dogma. The newness of science can be more fully appreciated by another development during Darwin's life: when Darwin began his famous voyage on the Beagle in 1831, the term scientist didn't even exist; it was only in 1883 that William Whewell coined the term [5].

These historical details underscore the recency of modern science, and strongly suggest at least one powerful reason why it took so long to take hold: people feared contradicting powerful religious dogma. But is that explanation fundamental, or is there a deeper level to this mystery? And why does opposition to certain scientific findings increase as supportive evidence does, as happened in the Galileo case, and as is happening even today in some areas, most notably evolution? Fundamental and retrospectively obvious discoveries are still made, and their apparent obviousness forces people to wonder how they remained undiscovered for so long. Many who fruitlessly prospected the same intellectual territories, but habitually overlooked the now-obvious riches are secularists and even self-described atheists.

Is it possible that conventionalism, rather than religion *per se*, is the more fundamental problem? We can't ignore such strong evidence—maybe not pointing away from religion so much as pointing toward more fundamental human limitations as ultimate motivations for persecution of ideas that catalyze social upheaval. When important truths lie long undiscovered, and we are seduced into wondering how so many could have been so blind for so long, we should take a moment to realize that a vast treasure of undiscovered truth still lies in plain view before us all. The now obvious wasn't at all obvious a short time ago, and the completely non-obvious will soon be obvious—that is, once someone has done the difficult work of overcoming the innate conventionalism of the human mind.

5.3 A Mind Lost in Time

The fact that science is so young has important implications for our future. Most importantly, it provides convincing testimony that human minds are not good at science. Some minds are better than others at science, but the basis for a substantially better future is the acknowledgment that the human mind in its current form is insufficient for certain critical challenges now facing humanity. Albert Einstein, who is considered one of the greatest scientists in history, remarked (during the year following the atomic devastation of Hiroshima and Nagasaki) that "a new type of thinking is essential if mankind is to survive and move toward higher levels" [6]. James Watson, the co-discoverer of the structure of DNA, directs characteristically blunt criticism at scientists, saying "most scientists are stupid." Watson explained: "Yes, I think that's a correct way of looking at it, because they don't see the future" [7]. Understanding the present well enough to predict the future with reasonable accuracy is an extremely important type of intelligence, and it contributes to good science. Nevertheless, Watson's relativism excuses the failings of better scientists. Again, humans are not good enough at science, and that means *all* humans. This point is sure to be contested, but alternative explanations are very weak or simply unacceptable.

Those who counter that some people are sufficiently good at science must confront the unavoidable ethical dilemma accompanying such a belief: they either don't believe science has the power to fix human problems and assuage suffering, or they don't care to assuage it.[1] Generalizing from the abundance of caring scientists we know leaves only one explanation consistent with all evidence: human minds as they currently exist are not capable of effecting our most desirable present and future. When we consider that our future depends fundamentally on our minds, both the challenges and the most efficient solution are made clear.

Here is a key question: *why* should we try to cope with modern, complex civilization, using brains provided by nature for use in a simpler time; brains that have been shaped and constrained by forces that are either irrelevant or quickly becoming so? For example, consider the expense of brains over evolutionary time. The human brain is very large for body size, relative to other species, and countless women have died in childbirth (and still do) as the size of the brain increased well beyond the typical ratio found in other species. Both fetal head size and the additional food energy required in the mother's diet ensured that *in utero* brains were under strict constraints that have become more relaxed.

Furthermore, the adult human brain is about 2 % of total body weight, but generally consumes more than 20 % of daily food energy intake. As a result, making a bigger brain has been very expensive over evolutionary time. Harvard anthropologist

---

[1] The argument that overall progress is slow because science is inevitably slow is a conventionalist fiction that conflates human inefficiencies with scientific ones. Consider the practice of science and engineering at the highest imaginable level (for argument, consider god-like abilities). We take it as given that a being with such abilities would be capable of assuaging most or all human suffering in short order.

Richard Wrangham has advanced the compelling hypothesis that fire was of primary importance in human evolution because cooking allowed a quantum leap in the amount of energy obtained from a given piece of food [8]. He suggests that this critical advance helped to launch a phase of rapid evolutionary change in the size and power of the brain. Several important elements needed to be in place in order to discover and exploit fire, but one of them was sufficient intelligence, and that type and level of intelligence was further amplified by the reliable domestication of fire.

This general strategy of developing and using technologies has ultimately leveraged existing intelligence through incrementally higher types and levels of intelligence over evolutionary time. Such "bootstrapping" has been selected for because there are reproductive rewards that accrue to an organism able to adapt quickly to new niches, or even able to create or modify existing niches to better suit their existing biological limits. These are essential features of what we think of as higher intelligence. Even though this process requires expensive fuel for nature's tinkering on the brain, sentient life's most metabolically costly organ, this expense reduces the cost of useful information. This inverse relationship must have been fundamental to the evolution of cognition, and it suggests a question: is there a point in the evolutionary process where useful information becomes so costly that the price of building a better brain is too high? The answer must be yes.

Even a large and powerful brain is confronted by challenges that are potentially rewarding, but for which optimal answers cannot be found soon or in the local environment. Even for countless simpler problems, the set of possible solutions is infinite and only some are practical and efficient. Random trial and error explorations of an infinitely large "solution space" will not often be rewarded. There are many types of information that might benefit us, but many are extremely expensive to both acquire and maintain. Given that brains are expensive, and that information can be both difficult to acquire yet extremely valuable for survival and reproduction, there will exist a constant tension—an unbridgeable gap—between what we have and what would benefit us more. UCLA anthropologist Rob Boyd and UC Davis evolutionary sociologist Pete Richerson have extended economic theory into the study of evolution and focus primarily on the acquisition of knowledge. Boyd and Richerson's "costly information hypothesis" is premised on the idea that when information is costly to acquire, it pays to rely upon cheaper ways of gaining information, and these are generally obtained through social interaction and instruction [9]. Note that their hypothesis is essentially just another way to say brains are expensive, except that they focus on the cost of information rather than the cost of the mindware (in this case, brains) needed to process that information.

In general, it is cheaper to learn from or mimic someone else's sequence of words, actions or expressions than to learn a complex behavior by experimentation. When information is dangerous, time-consuming, or difficult to acquire and process, learning by mimicking others will be selectively advantageous. Such a strategy for acquiring new information has obvious implications for adherence to convention, and for constraining innovation, including in the sciences. Boyd and Richerson have built a very solid formal foundation for this theory, and they make a compelling case that it explains many apparently maladaptive behaviors. As we consider the evolutionary tradeoffs that have shaped the human mind, and acknowledge that essentially all the evolutionary constraints and costs of building better brains and other thinking machines have declined substantially or disappeared, we are left to ask again, why should we continue to struggle to get by with brains mismatched to the complex world we now inhabit?

5.4 A Fundamental and General Solution

Typical proposals for reducing the impact of problems faced routinely by people in all parts of the world focus on treating symptoms rather than root causes. There is often no commonality of goals, and no sharing of resources produced for each of the litany of serious problems facing humanity today. In fact, the opposite is true: many strategies for solving disparate challenging problems compete for funding and attention. It would be highly beneficial for people to begin thinking more efficiently, cooperatively, and synergistically, and seriously consider more fundamental solutions that can be applied to problems more generally.

The most efficient and general solution to all human problems is to enhance our fundamental abilities to solve problems. A dizzying multitude of technologies have been developed for enhancing our physical selves and environments. Tools and techniques have been created to feed, clothe, and care for our material wants and needs. We have, with machines of human design, wrangled rivers and moved mountains; we routinely fly people around the globe and sometimes even into space; we have tapped the planet for its finite bounty, to suit our immediate desires. But this enhancement of humankind's physical abilities has expanded at a greater rate than our capacity to wield such power responsibly, and to foresee the long-term consequences. Only recently—only through this young mode of problem solving that we call science—has a realistic approach to enhancing our innermost selves become conceivable.

Increasing and refining human abilities to solve problems is not a new endeavor. Modifying the mind is a practice visible in every classroom around the world. The act of instruction originated before recorded history, and indeed, before humanity. Learning through traditional means physically changes the structure of the brain, but is slow and inefficient. A complete professional education, from primary school through college and then graduate school, is expected to take well over two decades. Education is the best technology currently available to alter human minds, but it is demonstrably too slow and too narrow to address and surmount the complex threats we face. Education is alteration, but it is not enhancement; it falls short of fundamentally augmenting the evolved potential or upper limits of the mind.

Many leading scientists and technologists recognize the fundamental importance of better problem-solving abilities and favor the pursuit of Artificial Intelligence (AI). We agree that AI is important and we regard some AI efforts as extensions of human minds. And we accept the general classification of natural and artificial intelligence under the umbrella term "mindware" [10]. However, the creation of "standalone"

AI that has its own interests and goals, potentially separate from those of humanity, is an uncertain proposition that has unsettled many futurists. A primary worry is that AI will view humans as short-sighted, irrational, and excessively aggressive, and it will arrive at the only possible logical deduction: extermination of humans in the name of self preservation. To circumvent such an outcome, AI might be created with an immutable friendly bias towards humans, or with an absolute dependency on human caretakers or symbionts. But these systematic constraints must be perfectly inviolable, which will depend completely on the mental capabilities of the systems' creators.

If AI is created carelessly, we agree that the probability of the AI doomsday scenario is substantially greater than zero. But we disagree with an exclusive focus on limiting the abilities or powers of AI to respond to the dangers to AI posed by humans, since humans are a more general threat that is no less real or serious outside the context of AI. Those who worry about the AI doomsday scenario, but who focus exclusively on the AI side of the equation, implicitly validate our belief that the limits of human minds are the more fundamental problem. This can be seen in the apparently paradoxical answer to the question of whether the human mind or AI is more unpredictable and dangerous to the human future. If we believe an AI would be a highly astute judge of risks, even an answer of AI betrays a belief in potentially catastrophic human mental limitations.

This logical form can be generalized universally to reduce the complexity of the landscape of vexing challenges and proposed solutions. Most relevant in the context of this essay contest, concerns about the future of humanity and civilizational risk[2] reduce to a more fundamental concern that our minds are insufficiently able to appreciate and/or handle the challenges before us. Better mindware is arguably the *only* technology capable of counteracting the myriad complex obstacles, problems, and threats facing humanity (including or especially those for which humanity played a contributing role), and better human minds are indispensable even to the pursuit of a general AI. Thus, better minds provide a truly fundamental and general solution, and to our knowledge, no other problem-solving approach is worthy of such a claim.

5.4.1 The Path to the New Mind

Some of the most threatening global problems have remained tenaciously intractable over the past decades, irrespective of national wealth and technological achievement. Even developed nations suffer from stubbornly stable levels of mental illness, poverty, homelessness, crime, and incarceration in otherwise increasingly wealthy economies. Many interventions have been tried, in an effort to reduce poverty and homelessness, including provisions of social services, food allowances, housing benefits,

---

[2] Civilizational risk is our preferred term for what many call existential risk. It establishes a lower limit for acceptable risk since we value civilization and neither individual nor group existence is threatened by many threats to civilization.

employment resources, various kinds of training and education for all age groups, so-called microloans and other loan guarantees, and so forth. But careful research shows that the primary driver of apparent cycles of social ills is the mind: mental health services improve social conditions, but improved social conditions do not improve mental health and functioning [11].

Mental health research and treatment represents a gateway to the unprecedented and uniquely important enhancement of human minds. Technologies spanning across the fields of genetics and genomics, synthetic biology, neuroimaging, brain-machine interfaces, and others are becoming increasingly powerful, with immediate applications for understanding and treating mental dysfunction and disease. However, these developments are relevant beyond treating mental illness. Given that even the most "normal" human mind is in many ways disabled by naturally imposed limitations, research focused initially on mental illness can provide entree to a more general research platform for mind engineering. This engineering provides a possible escape from outdated and destructive cognitive constructs, which produce and exacerbate human suffering and civilizational risk—but we must be very careful in the design and creation of new and better mindware.

It is essential to recognize that limits of even normal or high-functioning human minds are not only quantitative, e.g. processing speed or memory capacity; minds are also limited qualitatively in the kinds of biases they exhibit and types of errors they make. Daniel Kahneman's 2011 book *Thinking, Fast and Slow*, became an instant classic in human psychology and decision making [12]. In it he reviews a wide range of empirical tests of beliefs and behaviors, and concludes that people exhibit many biases including a "pervasive optimistic bias," which he says might be "the most significant of the cognitive biases." While such a bias might seem preferable to others, Dr. Kahneman says that it regularly results in unrealistic and costly decisions. Decades of research support Kahneman's claim that the optimistic bias is pervasive. In 1969, Boucher and Osgood suggested that languages have an inherent positive bias [13], and as of 2014 this hypothesis has been confirmed in all languages tested [14].

How can this be? How might evolution reward unrealism, ultimately producing a mind that creates an internal mental image that is discordant with external reality, and even with its own knowledge of itself? In a now famous foreword to the 1976 First Edition of Richard Dawkins' *The Selfish Gene*, evolutionary biologist Robert Trivers established a new perspective on how evolution shapes the mental realm. He wrote "If … deceit is fundamental to animal communication, then there must be strong selection to spot deception and this ought, in turn, to select for a degree of self-deception, rendering some facts and motives unconscious so as not to betray—by the subtle signs of self-knowledge—the deception being practiced" [15]. Harvard psychologist Steven Pinker suggests that this single sentence "might have the highest ratio of profundity to words in the history of the social sciences" [16]. Trivers' catalytic insight helps us to understand how evolutionary forces might create unrealistic and self-deceiving mental architectures, wherein unrealism isn't just a random or unselected trait—or a trait against which selection acts—but a purposely selected trait. Even prior to this important change in perspective, scientists in many areas had provided empirical evidence of the flaws of normal and even high-functioning minds.

And in the decades since, many psychologists like Kahneman have provided strong empirical support for this counterintuitive idea that evolutionary selection can favor varying kinds of unrealism—with excessive optimism being only one of many.

These theoretical and empirical revelations about how human minds actually function have profound implications for research and development of both natural and artificial intelligence, but these implications are widely unrecognized or underappreciated. Some have advocated enhancing human intelligence absent apparent concern about rationality or realism. Others have proposed the construction of a general purpose AI or artificial mindware that is based on the function—and in some cases even the physical architecture—of the human brain [17–19]. However, we are unaware of the realistic portrayal of human brain function, and its intrinsic biases and limitations in these proposals. In contrast to the common portrayal, the mind has not evolved to produce accurate internal representations of external reality, or even of its own internal processes and views. So, models or emulations of the brain as it exists will not and cannot produce a general-purpose, dispassionate, and realistic problem-solving mindware. A more likely product of such efforts is mindware possessing typical human faults, including routine unrealism and irrationality. What might be the outcome of empowering self-deceiving mindware with superhuman intelligence and powers of self improvement? One possibility is that it would improve itself on a trajectory of increased realism and avoid causing serious harm in pursuit of unreasonable goals, but we simply cannot predict what course it might take. We take a similarly cautious view of enhancing human intelligence across the existing spectrum of human (un)realism and emotional (in)stability.

These thought experiments highlight the importance of enhancing traits of existing minds in a preferred order, and in the creation of an AI with certain improvements relative to humans. Space constraints don't permit a thorough consideration of trait prioritization but two points are worth mentioning. First, evolutionary forces select for short-term reproduction over longer term sustainability; therefore, one challenge is to progressively de-emphasize short-term "band-aid" approaches to vexing problems, and to increasingly emphasize long-term approaches for growing and stabilizing civilization. Second, great care must be taken to establish a priority order even for preferred traits; there are few mental traits (maybe as few as one) that should be the initial focus of a trait prioritization plan. Bearing in mind these two points, consider near-term self-centered happiness and long-term rationality as two exemplary complex traits. Envision the enhancement of near-term happiness absent a minimum level of long-term rationality. A reasonable case can be made that such enhancement already exists in addictions to drugs, such as heroin or cocaine. Similarly, enhancement of intelligence absent rationality or certain other emotional stabilizers might be equally dangerous to long-term interests of self or others.

Our intention here is not to present—or even begin—an ordered list of preferred traits, but to catalyze discussion, research and development of better mindware. An important element of that effort is to focus on desirable traits neglected by or selected against by evolutionary processes. In that spirit, and in agreement with certain efforts already underway [20], we suggest that long-term rationality is a candidate for initial enhancement efforts. We believe this high-level trait embodies multiple narrower traits, including some consistently overshadowed throughout natural evolution by short-term self interests: empathy, group interest, quantitative long-term modeling and prediction, among others. One question in the pursuit of better mindware is "how will we produce mental traits that are beyond current human limits?" We can only offer the observation that the creation of "supernormal" traits obviously occurred routinely throughout evolutionary time, and the belief that such bootstrapping should not be beyond the reach of the best human science and engineering. At each successive step up the scale, supernormalcy will become the new normal, and so on into the future.

5.5 Comment And Summary

To answer the question posed by this essay contest, "How Should Humanity Steer the Future?", rather than provide a detailed plan, we argue that there is a single most-efficient overall focus on R&D of better mindware. We thank the many people who conceived, managed, and judged this essay contest, and we hope it provides a watershed moment in the discussion of civilizational risk. The submitted essays provide an excellent resource for advancing this discussion. The central recommendations of the essays reveal a typical propensity even among highly intelligent and educated people to treat secondary phenomena (symptoms) rather than root causes, validating one important pillars of our argument. We nevertheless concede that the outstanding prizewinning essays provide compelling reasons for immediate focus on a few critical areas in addition to a focus on mindware. But we are especially gratified to see that aside from our piece, some other fine entries—including the First Prize essay by Sabine Hossenfelder—focused on the most fundamental determinant of civilizational success or failure: human minds and other mindware. We are confident that our essay took Third Prize because of the superiority of the essays that finished ahead of ours, and not because our premise is unsound.

Minds are central; they are the foundation of humanity's past, its present, and its future. Human minds are the root cause of all problem-solving inefficiencies, but they are also the only creative engines capable of taking on each of these challenges, and of designing and building a better future. The evolution of the human mind allowed us to rise to a position of pre-eminence on our planet, but a rise to dominance in the past does not presage control over the future. As circumstances change dramatically, so must our thinking—and ability to think—to survive and thrive indefinitely into the future. All people—especially scientists and engineers—who are interested in building the best possible future must contribute to humanity's effort to design and build better mindware. This is the greatest-ever challenge in the history of humankind. Among the countless billions of species to ever inhabit planet Earth, it is only ours that has the unique privilege of taking this bold step. We owe it to our descendants that they should have more and better than we have, and that they *are* more and better than we are, yet they depend completely on us to rise to this challenge.

References

1. Paulson, T.: The lowdown on topsoil: it's disappearing. http://www.seattlepi.com/national/article/The-lowdown-on-topsoil-It-s-disappearing-1262214.php. Accessed 8 Apr 2014
2. Smith, A.: Future of Technology | Pew Research Center's Internet & American Life Project. http://www.pewinternet.org/2014/04/17/us-views-of-technology-and-the-future/. Accessed 19 Apr 2014
3. Armitage, S.J., et al.: The southern route out of Africa: evidence for an early expansion of modern humans into Arabia. Science **331**(6016), 453–456 (2011)
4. Scientific revolution—Wikipedia, the free encyclopedia. http://en.wikipedia.org/wiki/Scientific_revolution. Accessed 13 Feb 2015
5. Whewell, W.: (Stanford Encyclopedia of Philosophy). http://plato.stanford.edu/entries/whewell/. Accessed 18 Apr 2014
6. Einstein, A.: Albert Einstein—Wikiquote. Wikiquote (2014). http://en.wikiquote.org/wiki/Albert_Einstein Accessed 18 Apr 2014
7. Watson, J.: PBS—Scientific American Frontiers: The Gene Hunters: Resources: Transcript. (2014). http://www.pbs.org/saf/1202/resources/transcript.htm. Accessed 18 Apr 2014
8. Wrangham, R.: Catching Fire: How Cooking Made Us Human. Basic Books, New York (2009)
9. Richerson, P.J., Boyd, R.: Not by Genes Alone: How Culture Transformed Human Evolution. University of Chicago Press, Chicago (2005)
10. Rothblatt, M.A.: Virtually Human: The Promise—and the Peril—of Digital Immortality, 1st edn. St. Martin's Press, New York (2014)
11. Lund, C., et al.: Poverty and mental disorders: breaking the cycle in low-income and middle-income countries. Lancet **378**(9801), 1502–1514 (2011)
12. Kahneman, D.: Thinking, Fast and Slow, 1st edn. Farrar, Straus and Giroux, New York (2011)
13. Boucher, J., Osgood, C.E.: The pollyanna hypothesis. J. Verbal Learn. Verbal Behav. **8**(1), 1–8 (1969)
14. Dodds, P.S., et al.: Human language reveals a universal positivity bias. Proc. Natl. Acad. Sci. USA **112**(8), 2389–2394 (2015)
15. Dawkins, R.: The Selfish Gene. Oxford University Press, Oxford (1976)
16. Pinker, S.: Representations and decision rules in the theory of self-deception. Behav. Brain Sci. **34**(01), 35–37 (2011)
17. Kurzweil, R.: How to Create a Mind: The Secret of Human Thought Revealed. Viking, New York (2012)
18. Markram, H.: The blue brain project. Nat. Rev. **7**(2), 153–160 (2006)
19. Hawkins, J.: On Intelligence, 1st edn. Times Books, New York (2004)
20. The Long Now Foundation. http://longnow.org/. Accessed 2 Mar 2015

Chapter 6
A Participatory Future of Humanity

Dean Rickles

Well, here's another nice mess you've gotten me into.
Oliver Hardy to Stan Laurel (*Utopia* [1951])

Abstract History is rich with examples of humans expecting future generations to deal with their mess (be it debts, environmental impacts, or whatever); this can include people taking from their own future selves. The larger problems of humanity are, I argue, but scaled up versions of this same curious, irrational behaviour. Humanity's steering of the future must involve going beyond humanity in some sense. The solution I outline in this paper involves a modification of the everyday human stance towards future events and future selves. It involves a (practical, day-to-day) denial. Using a range of examples from physics, philosophy, neuroscience and psychology, ultimately advocates an intervention indicating how actions now are linked to future experiences and events, that agents themselves will have and influence by direct creation. This might seem blindingly obvious, yet the vast majority of humans act as if their lives are determined by the whims of the future. If humans fully realised how tightly bound they are to their future conscious life and experiences (and others'), and how that life and experience is a direct extension of their life and experience right now (so that actions can be seen to have direct consequences for their present selves: those that are experiencing right now), then they will be far more responsible in choosing their actions.

Like Laurel and Hardy in *Utopia* (their final movie), we are cast adrift on what could remain a beautiful island. In their case, it turns out to be a uranium rich resource. Greed-based madness ensues, in which 'Atoll K' is plundered, and a beautiful exis-

I would like to acknowledge the Australian Research Council for funding.

D. Rickles (✉)
Unit for History and Philosophy of Science, University of Sydney, Sydney, Australia
e-mail: dean.rickles@sydney.edu.au

tence ruined. Sound familiar? If we're not very careful as humans, this might also be our final performance...

What can be done to avoid a seemingly inevitable calamity—our *final century*, as Martin Rees [11] puts it? I fear a terrible end really is inevitable unless immediate drastic action is taken. That such an end is plausible can be seen by simply inspecting past complex societies and their collapse [16], combined with (1) the absence of any alteration in the more damaging human attributes and (2) a far greater aptitude for destruction. These attributes are virtually endemic on the Earth as a whole and are often connected with greed. Even alcohol, drug, food, gambling and other binges and addictions are small, local instances of a larger phenomenon: instantiation of actions now that will lead our (and other) future selves into misery. Procrastination clearly falls within this category: not performing some (less desirable) action *now* can lead a future self into abject misery and poverty. History is rich with examples of humans expecting future generations to deal with their mess (be it debts, environmental impacts, or whatever). The larger problems of humanity are but scaled up versions of this same curious, irrational behaviour. Humanity's steering of the future must involve going beyond humanity.[1]

The solution I outline in this paper involves a modification of the everyday human stance towards future events and selves. It involves a (practical, day-to-day) denial of fate and lack of control over future events along with a greater responsibility towards future selves. It ultimately advocates an intervention indicating how actions *now* are linked to future experiences and events, that *agents* themselves will have and influence by direct creation. This might seem blindingly obvious in reflective mode, yet the vast majority of humans act as if their lives are determined by the whims of the future. If humans fully realised how tightly bound they are to their future conscious life and experiences (and others'), and how that life and experience is a direct extension of their life and experience right now (so that actions can be seen to have direct consequences for their present selves: those that are experiencing right now), then they will be far more careful in choosing their actions.[2]

---

[1]I share with David Bohm the belief that (1) most of our lived reality (money, airplanes, class, national boundaries, etc.) is the result of human thought and (2) a kind of 'thought malfunction' is behind many of humanity's ills—see, e.g., [1]. I differ in that I find humanity to be very strongly characterised by such 'malfunctions' (Bohm was less cynical). Stuart Sutherland puts it well: "*Pace* Aristotle, it can be argued that irrational behaviour is the norm not the exception" ([15], p. ix). Hence, I believe that shifts in certain deep structures of human thought are required to resolve the problems we face.

[2]Take the simple act of drinking too much alcohol. Everyone 'knows' that a hangover will result in the future. Yet they proceed anyway, passing the bad experiences onto their future selves. They wouldn't choose to give their present self a hangover, of course. But the difference between this 'present self' (what is being experienced right now) and a future self (what their present self *will* experience) is hardly anything at all. There *will be* a present self that will have the hangover, and this self is (philosophical conundrums aside) no different from the one that is now throwing back the alcohol: one's future selves *will be* present selves soon enough! Why think it is OK to pass bad experiences onto a future self, given that you are essentially the same self (in the sense that you will be the one experiencing it)? There is no rational reason, and yet many of us do it, over and over again, passing the buck to our poor, suffering future selves. Whatever one makes of this scenario, I

In other words, presentist thinking is the problem. Thinking that all that matters is what happens now is the problem. Spatiotemporally local thinking is the problem: events that are distant in space[3] or time do not play a strong enough role in human decision making. This (primitive) instinct must be transcended and a habitual eye to future selves somehow enforced.

6.1 The Humean Condition

In his rather remarkable essay, "On the Populousness of Ancient Nations" (1777: reprinted in [3]), David Hume argues that human nature has remained pretty much constant over time, as has a penchant for complaining about one's present times in favour of the past! I wonder, however, whether we are now genuinely justified in complaining about the present state of humanity? While I agree, as mentioned above, that there is a constancy in human nature, there isn't a constancy in the way humans operate in the world. New theories and technologies can radically alter the playing field. What has changed from past problem societies is the number of ways in which we can (or in which we might) conceivably be wiped out—though, curiously, the count it is highly dependent on our current theories and, as such, subject to revision as much as those theories are (e.g. germ theory was not known 200 years ago, and so there was no known risk associated with germ-specific pandemics): as we develop more knowledge about the way the world works, we must revise our assessments about the state of humanity, and the extent to which our predicament is dire. But, in any case, the question 'How should humanity steer the future?' presupposes that the current trajectory is not a good one and that humans can and do play a guiding role. Given this, let's state some facts about humans and the world:

---

think it points to the fact that struggling with problems of the future of humanity (and local versions thereof, such as procrastination) demands that we think deeply about such philosophical issues as the nature of selves over time and our responsibilities to selves at other times.

[3] That is, in general, somebody in need of help within my personal space will command more urgent attention than one that is spatially distant, despite the fact that there exist spatially distant persons more in need. The issue boils down, in part, to what one feels able to influence. If one felt able to influence the spatially distant needy, one would be more likely to do so. Likewise with temporal distance (though, of course, only in a preferred direction in this case, due to the asymmetry of influence, in the absence of time machines!)—in fact, I think we should (and often do) have moral empathy in a temporally neutral way, as when we mourn the victims of Auschwitz or those killed in the eruption of Mount Vesuvius in AD79, for example. Hence, one needs some way of linking people up more robustly with their future selves so that their feelings of responsibility towards them is increased. Of course, this temporal myopia is not true of all actions (savings, pensions, quitting smoking, and diets are straightforward counterexamples), but I think it is true of the majority, and even for those able to delay gratification it is hard to maintain a consistent standpoint according
(Footnote 3 continued)
to which one's future selves are equally as important as a (current) present self—I might add that the widespread prevalence of *financial credit systems* in human society is a strong indication that a temporal myopia (passing responsibility to future persons) is at our core.

1. The planet is in a dire state: socially, politically, economically, and naturally.[4]
2. To an undeniably large extent, humans are responsible for the dire state of the planet.

Human thought and action has caused much of the damage. These actions were caused by their psychological profiles. Psychological profiles can be modified and behaviour patterns can be modified. This has a ring of something horrifying, but the alternative (extinction) is more horrifying.

Many of the problems are rooted in the fact that people believe that the way things are at any time are how they *have* to be. They find it difficult to question the (or their particular) status quo. Conflict arises when there is difference of opinion or belief, given this rigidity in thinking.[5] Such differences are of course rife. An obvious strategy is to push for greater uniformity of opinions and beliefs. I think this will be part of a solution to humanity's problems (cooperation is of course easier within a shared framework of beliefs), but the problem is: whose opinions and beliefs? Consensus is difficult in the simplest of scenarios. However, it would be nice to think that one day people will be guided by the best available evidence, and nothing else... Unfortunately, I think a uniform, evidence-based world is too hard to engineer (ethically as well as practically). So how else might we tackle the issues?

6.2 Trans²humanism: Why Technology Just Isn't Enough

Stephen Hawking recently argued that spreading to other parts of the Universe constitutes the only plausible survival strategy of humans.[6] This is entirely wrong headed. It runs away from a central problem: what good is it spreading humanity across the universe, if in the process we spread the same destructive traits? In fact, I'd consider this to be an immoral act: the survival of humanity (the end) in its present condition does not justify Hawking's proposal (the means). Neither technological solutions, nor advanced technology, are enough for this reason. What we have here is a basic problem of human nature: it tends to think in terms of 'now,' with an improper sense of how actions now create future nows (which include future experiences). This can be seen in the patterns of self-destructive behaviour (both individual and societal) repeated again and again, resulting in the repeated collapse of complex societies.

---

[4]It isn't all bad, of course. In, say, the last century: mortality rates have declined; education rates have increased; social justice and equality appear to have improved enormously; and overall quality of life appears to be much improved. However, even these silver linings still envelop dark clouds thanks to population calamities and financial/resource/environment decimation and pollution.

[5]Religious beliefs are certainly amongst the most rigid (since they are not evidence-based and therefore harder to update). I can't help thinking that a belief system that preaches entitlement to the Earth and all its contents can't be a good thing—as others have noted, religious belief appears to be correlated with a general neglect of the Earth and it's non-human animals. Even more dangerous is the notion that this Earth might be little more than a 'staging area' for an afterlife: why would one care about its future state if one truly believed this?

[6]http://bigthink.com/big-think-tv/stephen-hawking-look-up-at-the-stars-not-down-at-your-feet.

The way we view ourselves as situated in space and time is, I shall argue, in large part responsible for the pattern: collapses are viewed as things that happen to past civilisations. Yet they were formed by an aggregation of (avoidable) once present actions.

Am I then simply proposing transhumanism? Not quite. Transhumanism generally involves *transcending* the human body.[7] But, again, what good is this if the same human *nature* (in the wider sense) lies behind it? Transhumanism ought to involve a transcendence of human nature since that is the problem: trans^{2}humanism. Human nature is possession and greed. Possession and greed often stem from an overly local mode of thinking: to want some item right now, regardless of the future implications, and regardless of how similar occurrences have gone in the past. Human nature is the tragedy of the commons, taking everything for oneself and leaving others to deal with the implications. These are attributes worth transcending.

6.3   QBism and the Participatory Universe

Personal experience (the stuff that matters most to humans) has been eroded from science. Humans have been made to feel small and insignificant—this is plain to see in the 'principle of mediocrity,' *viz* humanity does not occupy a special place in the Universe. As in physics, human decision making, especially towards events in the far future, seems to eliminate the subjective point of view to a remarkable degree. It matches the scientific objectivism in which the scientist, as Schrödinger put it, "simplifies his problem of understanding Nature by disregarding or cutting out of the picture to be constructed himself, his own personality, the subject of cognizance" ([13], p. 92). This is a problem, for it is *us* that will be faced with the stages of any 'action sequences.' A transformation is needed in which present experience is also viewed as encoding a blueprint of the future. Enter QBism.

QBism has its roots in a personalist, subjective Bayesian account of quantum probability along the lines of de Finetti. What really matters is that this probability forms a 'coherent' (consistent) framework. Strictly speaking, however, probability is not something that exists 'out there,' in the world: it is something assigned by an *agent*. This quite naturally affects the interpretation of the quantum mechanical wave-function, Ψ, since that is usually taken to represent probabilities for outcomes of measurements. Indeed, quantum mechanics is usually seen to be a *theory of probabilities of outcomes*. If probability does not have any objective existence, then, according to QBism, neither do quantum states. Rather, as with probabilities, they too reflect something subjective.

The central feature of QBism that I wish to draw upon is the notion that measurements are 'experience-eliciting' procedures by agents: they refer to the subject's

---

[7]It is, after all, often expressed as a transcending of "human nature"; but this latter expression, to my mind, denotes a moral aspect to that is missing from the majority of accounts of transhumanism which tend to focus on morphological freedom alone.

point of view, just like the probabilities. An agent assigns wave-functions to physical systems based on past experience and any other relevant facts. Quantum mechanics is then the tool used to make inferences about the agent's future experiences. Such experience (once realised, whatever it might be) is then fed back into the agent's store of experience, which might lead to an updated wave-function assignment, which would lead to a revision in the experiences one can expect to have in future interactions with the system. And so on... There is a sense in which agency is at the root of measurement outcomes: it is a creative process. This subjectivity—putting agents at the heart of world-building—can play a vital role in linking us to our shared future. As Chris Fuchs so nicely expresses it: "our actions matter indelibly for the rest of the universe," for quantum mechanics "signals the world's plasticity ... [w]ith every quantum measurement set by an experimenter's free will, the world is shaped just a little as it participates in a kind of moment of birth" ([12], p. 172). This strikes me as the kind of mindset that could evade the presentist, fatalist thinking that is damaging our world. But, note, that I do not mean to suggest that QBism is itself a strategy for dealing with the problems; simply that there are approaches to fundamental physics that merge well with what I take to be a healthy view of the world and our place in it: subjects (= 'us') are part of the world. I take this nonetheless to be directly relevant for the present topic since it shows that physics is compatible with a more 'agency-based' (participatory) view.

6.4 Time, Selves, and Humanity

Ironically, what often isn't discussed in these 'future of humanity' contexts is the ontological status of the future, and its human occupants. We get so bogged down in details of how to avoid calamities that we forget that here are deep philosophical issues involved. But depending on the nature of the future, there will be very different outcomes and very different strategies that ought to be recommended. For example, there is an unavoidable conscious sense that the present moment contains all that exists, and therefore that present selves should be given complete attention. A similar problem affects spatial distances: if we aren't in direct contact with some problem (say, famine in Africa), then we are not so disturbed by it as if it were right in front of us, in our conscious experiences. But why should spatial proximity matter in such cases? Though it is often supposed that the folk concepts of space and time are quite distinct, here we see a similarity: the mind relegates any problem that is not nearby (in space or time) to 'unreality.'

To a certain extent, global news and internet have closed the spatial gap, but the temporal gap (the greater problem when considering the future of humanity) seems surely unbridgeable? However, I think the temporal case is an example of a more direct connection, and the above discussion of the Qbist interpretation reveals in a small way how that might make sense. After all, the spatially distant selves are genuinely distinct individuals, while our future selves are *us*.[8]

It is interesting to note that the default position for far-future events of optimists (don't have to do anything about it now) is similar in outcome to pessimists (*can't* do anything about it now).[9] Which relegates action to future others: *they'll* get us out of this mess later on. But without some reconfiguration of minds, the future persons will be in the same situation as us: pointing to yet more future persons, who are by then going to be in an even greater mess, and so on! Of course, I can't append *ad infinitum* here because there is a terminus to this passing on of responsibility: annihilation! Despite all of the dangers we are told about, humans get pulled into to one of two positions:

1. The future simply isn't real (so what can I do about it).
2. The future is already real (so what can I do about it).

Quantum mechanics gives some creativity to humans, as we have seen. They can carve reality. If humans had the belief that they were involved in the creation of reality, they would be more committed to their future and forge a third way. If one is aware that each action is creating the future, and creating a future self and a habitat for that future self, then one is liable to be more committed to making that future the best possible. In this case, the direct link between now and the future is bridged: a world of humans that are constantly aware of carving out their future selves and experiences.

Interestingly, in a different context David Mermin [7] has also drawn attention to links between philosophy of time and Qbist thinking. He points out that the 'obvious' existence (to a subject) of a Now (a local present) has standardly been viewed as opaque to physics. Mermin, as with other Qbists, argues that the present moment can be incorporated into a physical description by invoking Qbist thinking: subjective experience, (usually excluded from physical science), is the fundamental thing, and, as he argues, spacetime is an inference (an abstraction) from this. If we do not think of future experiences as given *sub specie aeternitatis* then we are more closely involved with them: we have some control over them.[10]

---

[8]Derek Parfit [9] has famously discussed the irrationality of this 'near-future bias' (which he describes by means of a character 'Proximus,' who has a lot in common with most humans), in favour of temporal *neutrality*, whereby location in time is irrelevant from a moral standpoint (using a similar comparison with spatial relations)—though I have framed things in terms of one's personal future experience (which runs counter to Parfit's principle of a more impersonal, outward-looking moral stance).

[9]Studies suggest a temporal bias whereby optimism about the future grows with the temporal distance from the present [6]. Economists model this phenomenon by the method of 'hyperbolic discounting' (which is known to be an inconsistent scheme, yet models human behaviour far better than any consistent model [17]).

[10]Whether there 'really is' a block universe or not does not matter from this point of view: all that matters is our *stance*. Even in a block view there is a role for local creation of one's future experiences. Note that I am not seeking to defend QBism here either, but only using it as an example of how sense can be made of a participatory scheme in which we play a kind of co-creative role in making the Universe what it will *become*.

This brings us to a potential conflict between physics in the Qbist vein and a desire to take a 'long view' of time—an objection raised by Laurence Hitterdale. QBism as discussed by Mermin puts the Now of an agent centre-stage. That is, the subjectivity (the fundamentality of experience) of the Qbist view might force us to think in purely presentist terms since we only ever experience a present moment. Given this, there is surely no alternative but to be temporally myopic? But this rests on a mistake. The fact that subjective experience is of a present moment is of course behind much of the temporal myopia we experience (and related biases). I agree that we do indeed focus on "present experience: here and now" (indeed, my argument is based on the idea that this is the root of many of humanity's problems). But the elimination of the block picture doesn't *justify* the behaviour (I'm not sure how it would for the reason given above: the present is not an unchanging present). One can still be concerned about present subjective experiences (rather than experience) in the knowledge that *future* subjective present experiences *will* be had, and can be good or bad depending on how you act in the here and now, and they will happen to *you*. The fact that there *isn't* a block picture associated with QBism means that there is a sense of being able to have a say in those future experiences. The point is that how one acts in this present moment is actively linked to the following present moment (experience). By showing how your future selves and experiences (with early training to avoid instant gratification for example) are a direct result of your actions we might be able to get beyond the biases.

6.5 Out with the Amygdala!

It's all very well saying that these curious quantum mechanical and high-falutin' philosophical ideas can have profound applications: but how do they fit the real world? To get some idea of how they might be realised, let's consider the Stanford Marshmallow experiment, led by Walter Mischel [8].[11] Here children were offered two options:

1. Take one marshmallow right now.
2. Wait 15 min and get two marshmallows.

Not surprisingly, there were differences partitioning the experimental population into two groups. There's controversy over just what the differences reveal—some children might be fine with one; some children might be adapted to an unreliable environment (leading them not to expect the second marshmallow). However, it seems that in some cases *delayed gratification* is playing a role. This concept clearly involves *time* in a fundamental way. One is delaying some immediate pleasurable outcome so that one's future self is even better off. One is actively engaged in *constructing* a happier

---

[11] See Joachim de Posada, "Don't Eat the Marshmallow!", for a review of this classic experiment: http://www.ted.com/talks/joachim_de_posada_says_don_t_eat_the_marshmallow_yet.

future self. There has to be some underlying mechanism binding the delay group's present selves more closely to their future selves.[12]

I have been suggesting that human motivation and decision-making is at the root of humanity's problems. Humanity has an impulse control disorder. It plays the short game ('hyperbolic discounting'). It's the same lack of self-control that is responsible for impulse buying, binging, etc., and is an emergent manifestation of the biology of addiction. Understanding these can point to way out of these problems. The title of this section is not meant to be taken too seriously: the amygdala is part and parcel of the learning mechanism (since it is implicated in the [negative and positive] reward system: [5]) and needs to be *used* rather than removed! It is not unlikely that the same mechanisms lying behind addictive behaviours is also behind the irrational behaviours associated with procrastination: the immediate (counterproductive) response (*not* doing the work one is supposed to be doing), triggers the same kind of reward that smoking a cigarette does for an addict. If this neurocircuitry could be intervened in (by therapy, teaching, or even biomedical intervention as [10] argue), then one might isolate the mechanism responsible for human self-destructiveness.[13]

There is a relatively simple neurobiological basis for such apparently aberrant behaviour. The brain rewards pleasurable actions, which increases the likelihood of *repeating* the action—and procrastination is more pleasurable than hard work! Repeating the action also decreases the pleasure provided (i.e. reduces dopamine production), which demands more of the action, and so on. Linking this back to our theme, the idea is that the more one procrastinates, the more one procrastinates. Since global problems (such as repeated collapse of societies) emerge from these small local events, an intervention at this level could have dramatic emergent consequences. Our reward circuitry needs to be reconfigured in the light of changes in human existence. It is adapted to a very different (more primitive) kind of existence.[14] The good thing

---

[12]Mischel argued that increased capacity for delaying correlated with decreased behavioural problems and increased intelligence in later years—this seems entirely unsurprising to me, but still highly relevant from the point of view of this essay competition. Note that Mischel's book *The Marshmallow Test: Mastering Self-Control* was published shortly after this essay was submitted for the FQXi essay competition (Little, Brown and Company, 2014). This contains several strategies for countering self-gratification, some of which are suggested here.

[13]These connections are clearly testable by fMRI techniques which would reveal a similar pattern of oxygen utilisation in specific areas (e.g. activation in the orbitofrontal cortex, prefrontal cortex, anterior cingulate, extended amygdala, and the ventral striatum). It is possible to locate (initially by gene knockout) specific genetic markers for susceptibility to instant-gratification type thinking, just as one might locate genetic markers responsible for addictive behaviours [2]. In addition, certain experiments performed on mice to knock out specific receptors believed to be implicated in addiction have led to a reduction in addictive, self-destructive behaviours.

[14]The existence of temporal biases are well known in psychology, and in most cases a good evolutionary reason can be found that would have benefitted our ancestors, but can backfire on us. We need to be transhuman in the sense of transcending these kinds of primitive instant reward (where's the next meal coming from?) type processes. So much of the stupid behaviour in the world is caused by the same kinds of reactions that our distant ancestors employed which were useful for them, but are leading us to ruin. Merely explaining that and why we have these various mechanisms leading to such temporal biases does not *justify* them, and if we have the power to eliminate them to insure survival of the species, and hopefully make us better beings in the process, then we ought to do so.

is, the reward system is highly persistent once some pathway has been established. If one can set up positive pathways for future-based thinking, then they are likely to be fixed in the hippocampus and become habitual.

To finish, I might throw in a wild hypothesis: Individuals with Asperger's syndrome (or a certain form of autism spectrum disorder, according to DSM-V) seem to have 'amygdala deficits' that match remarkably closely what I have been viewing as *positive* characteristics, in terms of enhancing our chances of survival. They have an ability to intensely focus on tasks and engage in goal-directed (i.e. future-pointing) behaviours. They often seem better at regulating their impulses and not being pulled into crowd scenarios. Their social brain 'deficits' confer a humanity-level advantage. Perhaps they are better adapted for the kind of world that requires more rational thinking? Rather than acting on fast-system information (such as facial expressions), so crucial to our ancestors, they operate on slower inferential systems.[15]

This is slightly tongue in cheek, of course. But it is clearly true that there are fairly large individual differences in impulse control and goal-directed behaviours, and these translate into an ability or inability to plan rationally and choose actions that do not necessarily generate immediate pleasure but serve to either generate beneficial future outcomes, or an avoidance of negative future outcomes. By focusing on these individuals (on where the differences lie), I feel we could go a long way to resolve some of humanity's worst excesses. Moreover, it needn't lead to a world in which humanity has been totally eradicated.[16] Combined with a perspective shift loosely informed by Qbist-type stances, and non-presentist thinking, we can allow humanity to steer itself out of danger—though ultimately, this viewpoint suggests that 'steering' is the wrong metaphor, since that implies a fixed, ready-made terrain *through which one steers*: I prefer the metaphor of *improvising* the future.

References

1. Bohm, D., Edwards, M.: Changing Consciousness: Exploring the Hidden Source of the Social, Political, and Environmental Crises Facing Our World. Harper, San Francisco (1978)

---

[15]Whereas neurotypical subjects will have activation in the amygdala when, e.g., making inferences on the basis of facial expressions, someone with ASD will instead have activation in the frontotemporal regions. Similarly, one finds a failure to imitate, especially where imitation is not goal-oriented [4]. These are usually labeled 'deficits'. However, again there is a sense in which an advantage is conferred. In a standard test, a therapist will perform an irrelevant ritual (say, tapping three times on a box) before removing an item from the box, before asking the subject to do the same. Neurotypicals perform the irrelevant moves, while those with ASD do not, instead completing the task in the most efficient way. It is true that such behaviours often lead to dysfunction in society (given the neurotypical dominance), however, it is also clear that the hyper-rationality such cases display could bestow advantages. The propensity for neurotypical humans to imitate also points towards a strategy for implementing new modes of living via a concerted shift in social norms, that will be widely adopted given the mirroring mechanism (cf. [10], p. 102).

[16]As posthumanists suggest, and as Roger Scruton objects to, for "[w]hy should we be working for a future in which creatures like us won't exist?" ([14], p. 230).

2. Crabbe, J.C.: Genetic contributions to addiction. Annu. Rev. Psychol. **53**, 435–462 (2002)
3. Hume, D.: In: Rotwein, E. (ed.) Transaction Publishers (2007)
4. Ingersoll, B.: The social role of imitation in autism: implications for the treatment of imitation deficits. Infants Young Child. **21**(2), 107–119 (2008)
5. Koob, G.F., Le Moal, M.: Neurobiology of Addiction. Academic Press (2005)
6. Lee, E. and Yoon, Y.: (2010) The effects of regulatory focus and past performance information on temporal bias of future optimism. KAIST Business School Working Paper Series No. 2010-009: http://papers.ssrn.com/sol3/papers.cfm?abstract_id=1707222
7. Mermin, N.D.: Physics: QBism puts the scientist back into science. Nature **507**(7493), 421–423 (2014)
8. Mischel, W., Ebbesen, E.B., Zeiss, A.R.: Cognitive and attentional mechanisms in delay of gratification. J. Personal. Soc. Psychol. **21**(2), 204–218 (1972)
9. Parfit, D.: Reasons and Persons. Oxford University Press, Oxford (1986)
10. Persson, I. and Savulescu, J.: Unfit for the Future: The Need for Moral Enhancement. Oxford University Press, Oxford (2012)
11. Rees, M.: Our Final Century: Will the Human Race Survive the Twenty-First Century? Heinemann (2004)
12. Schlosshauer, M.: Elegance and Enigma: The Quantum Interviews. Springer, Berlin (2011)
13. Schrödinger, E.: Nature and the Greeks. Cambridge University Press, Cambridge (1954)
14. Scruton, R.: The Uses of Pessimism and the Danger of False Hope. Atlantic Books, London (2010)
15. Sutherland, S.: Irrationality. Pinter and Martin (1992)
16. Tainter, J.A.: The Collapse of Complex Societies. Cambridge University Press, Cambridge (1990)
17. Thaler, R.: Some empirical evidence on dynamic inconsistency. Econ. Lett. **8**(3), 201–207 (1981)

Chapter 7
The Cartography of the Future: Recovering Utopia for the 21st Century

Rick Searle

> *A map of the world that does not include Utopia is not worth*
> *even glancing at, for it leaves out the one country at which*
> *Humanity is always landing. And when Humanity lands there, it*
> *looks out, and, seeing a better country, sets sail. Progress is the*
> *realization of Utopias.*
>
> Oscar Wilde, The Soul of Man under Socialism

The above quote from Oscar Wilde expresses a sentiment largely alien to the early 21st century. We really don't believe in Utopias anymore, or, if we do, associate them with the kinds of political violence found in the ideological movements that haunted the first half of the 20th century, state communism and Nazism, especially [1]. The road to Utopia, one suspects, leads to its opposite, to dystopia, visions of which are now all the rage [2].

Yet, we are as likely to ridicule Wilde's sentiment as we are to fear it. To characterize the views of someone as "utopian" is to call into question their very seriousness, to accuse them, in some sense, of being a fool. Utopians in this reading are either dangerous political fanatics or incurably naive, and perhaps in some cases even both. Wilde would look in vain to find Utopia on our maps.

In leaving Utopia unexplored we are abandoning a way of thinking with a very ancient pedigree. Human beings have been dreaming up perfect societies probably since we started living in cities, though, the Utopian idea was probably properly born only with Plato and his ideal societies as presented in works *The Laws,* and especially, of course, *The Republic.*

Imagining Utopia was one of the primary ways we have expanded our moral imagination. The Kallipolis of Plato's *Republic* did away with wars of imperial

R. Searle (✉)
725 East Freedom Ave, Apt # 1, Burnham, PA 17009, USA
e-mail: rsearle.searle@gmail.com

© Springer International Publishing Switzerland 2016
A. Aguirre et al. (eds.), *How Should Humanity Steer the Future?,*
The Frontiers Collection, DOI 10.1007/978-3-319-20717-9_7

expansion, established laws of war, freed slaves and gave women an equal place in society [3]. In the golden age of literary Utopias, from the 16th through the 19th century, authors and social reformers used ideal societies imagined and attempted in the real world to push forward social and intellectual reform [4]. Thomas More's famous *Utopia* was a less than veiled critique of nascent capitalism, and the corruption and militarism of early modern Europe [5]. Francis Bacon helped spark the scientific revolution with his *New Atlantis* seeing the purpose of the new science as a project of Christian charity, "the relief of man's estate" [6]. Social reformers who used small utopian communities to test their ideas were a common feature of the 19th century. With some attempting to discover ways capitalism might be made humane, such as those created by Robert Owen [7], while others were among the first to experiment with the abolition of chattel slavery, and gender equality [8].

Whole reform movements were born out of the new Utopian science- fiction created in the later 19th century, especially Edward Bellamy's *Looking Backward: 2000–1887*. Indeed, *Looking Backward* could be said to represent a turning point in the history of Utopia. Not only was his futuristic romance one of the first works of science-fiction, it had a huge effect on the public imagination. The third best selling work of fiction ever, *Looking Backward* sparked discussion clubs among the middle classes, actual Utopian communities, and was a source of inspiration for real world revolutionaries like Vladimir Lenin [9].

Perhaps more importantly it was a version of Utopia that would be impossible without technological progress to support the reconfigured social world it imagined. In some ways Bellamy might be thought of as a transitional figure in the Utopian tradition, signaling its long-term move away from values and towards dependence on technology, a move that would turn Utopia into both ideology and science-fiction.

Before the 19th century, Utopia, whether it was conceived as a blue-print for a new society, or merely as a critique of an existing social order, grew out of the desire to live in a world that better matched human values. Utopia was a society in a state of peace, with freedom from want, the absence of oppression and a myriad of other things that human beings had wished for at least since the time they had begun to live in cities. Even Francis Bacon whose *New Atlantis* was based on the growth of scientific knowledge saw his Utopia as a recovery of what was once our natural state [6].

We lost Utopia in this ancient sense once it came to be associated with a certain view of the future, a change in our relationship with time that came about because of the explosive growth of our knowledge and technological prowess.

Human beings are unique in our awareness of our extension across time. It is language that gives human beings a capacity neither other animals nor machines possess to be aware of the present as a continuum of the past and the future [10]. The past is essential to our existence and sense of ourselves and yet remains stubbornly outside of our control. It is only towards the future that our freedom has real meaning [11].

It is perhaps difficult for us to realize the idea that the future will be fundamentally different from the past is a relatively recent realization, though, as with seemingly everything else the ancient Greeks had hints of this. Empires might rise and fall and the end of the world would someday come [12], but for the majority of human beings day-to-day living would remain mind numbingly the same. What would break this cycle was the industrial revolution, which not only radically transformed human life, but promised, through unrelenting technological advancement, to continuously transform the world out into a now infinite future [13].

Both science-fiction and the ideological movements that came to supplant older versions of Utopia in the 19th and 20th centuries all grasped this new sense of the future and linked themselves to some notion of forward development, to progress, where the later stage in history was more advanced than the one that preceded it. Darwin's discovery of evolution itself seemed to give scientific justification for the theories of historical development espoused by the newly born science-fiction, and ideological movements.

Yet in thinking this progress was somehow the inevitable consequence of historical or natural laws we in some sense surrendered our own control over it. It is almost as if the minute we discovered that the future could be different from the past we latched onto a way in which our freedom over deciding what this future would look like could be minimized. Perhaps this was because the new technological change grew out of the success of the deterministic worldview of the physical sciences. The ultimate ambition of this deterministic philosophy was never better stated than by Pierre Simon Laplace:

> We may regard the present state of the universe as the effect of its past and the cause of its future. An intellect which at a certain moment would know all forces that set nature in motion, and all positions of all items of which nature is composed, if this intellect were also vast enough to submit these data to analysis, it would embrace in a single formula the movements of the greatest bodies of the universe and those of the tiniest atom; for such an intellect nothing would be uncertain and the future just like the past would be present before its eyes [14].

The idea of Utopia would henceforth rise or fall with the technological and corresponding historical determinism that the success of the physical science had aroused. It proved to be the case that a great deal of violence needs to be done to human beings and society in order to make them fit into the kinds deterministic reductions of the world that were inspired by Newtonian Physics and linked to technological progress, a tragedy we came to associate with the Utopian imagination itself.

Karl Popper was essentially right- the Utopias sought by the totalitarian movements, Nazism and Stalinism, were dangerous precisely because they treated human beings like Newtonian billiard balls or machines covered in flesh [15]. Where he was wrong was in projecting backwards into the whole history of Utopian thought the seeds of 20th century totalitarianism. The totalitarian movements were distinct from the Utopias that preceded them in that they tried to reinterpret history in light of the determinism of Newtonian physics. Their adherents believed not only that the future was determined, but that they knew the ultimate destination.

Many remain mesmerized by this conflated history that ties together the totalitarian movements and earlier utopianism, and continue to see in any discussion of Utopia the threat of violence and tyranny [16].

Yet it would be incorrect to think that the distortions of applying the determinism of classical physics applied to human society have been limited to the totalitarian movements of the last century. The physicist Lee Smolin has pointed out that the problem with contemporary free market economics isn't that it relies too much on quantitative models, but that its quantitative models built around concepts such as "market equilibrium" are based on a simplified version of science where the future was considered determined rather than open. By thinking the future is determined Smolin thinks we have surrendered our freedom in regards to it [11].

In yet another case of influence both fundamentalists and the narrow minded intolerant brand of new atheism they have inspired spring from the same determinist minded source [17].

Still, no social version of determinism is more important than technological determinism. Almost all other forms of determinism find their roots in the idea of advancing science and technology, and both remain the primary drivers of change in our world. Getting the question of technological evolution right will likely mean getting the future right.

Technological determinism can run both ways, but I will confront the stronger side of the argument. The case that technological evolution is leading to positive rather than negative outcomes is simply a better one than the reverse. Take any social measure you like such as longevity, child mortality, height, per capita income or level of societal violence, including war, and life is incomparably better after the industrial revolution than before [18, 21, 22].

The positive argument also has some pretty strong social forces behind it. Technological advancement is supported by rising groups, from the increasingly economically prominent technology companies who have overthrown or are challenging older rust and paper belt elites across perhaps all major economic sectors [19] to the push for technological advancement from the world's rival militaries [20].

Unlike most other forms of determinism that flowered in the 19th century and 20th centuries progressive technological determinism continues to have legs. A technorati semi-royalty such as the co-founder of *Wired Magazine*, Kevin Kelly, persists in making sincere and solid arguments, not only that there is a progressive direction to technological advancement, but will go so far as to suggest technology itself "wants" such an outcome [21]. Kelly is in good company, with other popular thinkers such as the founder of the X-prize, Peter Diamandis [22] along with other entrepreneurs and thinkers cranking out wildly popular books, the most famous of which is Ray Kurzweil, now Director of Engineering at Google, whose intellectual nuance consists of admitting that technological progress could lead us either to individual immortality or the destruction of our species [23].

The application of technology as the primary way to address our social problems has risen in tandem with a decline in our faith in political processes, in the ability of policy makers to effectively guide modernity. "Technological solutionism" as Evgeny Morozov calls it, is based the assumption that the majority of problems in human society are a matter of engineering, and has replaced politics as the default mode we use to address social ills [24].

The problem with the view that technological evolution has overall been incredibly positive for mankind is not that it is false. It is that the historical window it uses is far too narrow, and that such a view does not take into account the extremely contingent nature of our history so far. Properly speaking, technological civilization is only a little over two centuries old. Even if one shoves the window open to encompass the entire period starting with the creation of agricultural societies the period in which human beings lived in "technological" societies would make up a mere 5–10% of the history of our species [25]. Wherever we look in the heavens, ours remains the one and only test case of whether a technological civilization can survive over the long haul, and the long haul measured in millions or billions of years is very long indeed [26].

There is also the matter of our sheer luck. The story of progress looked very different at the height of the Cold War when it seemed like we might very likely blow ourselves up. In that era an insightful piece of fiction that dealt with our quest for knowledge, Walter M. Miller, Jrs' 1959 *A canticle for Leibowitz* presented the history of human knowledge and technology not as progressive but as an endless cycle of self-destruction and rebirth [27].

We should not assume that avoiding Armageddon was a pre-determined thing, for we came very close to it more than once [20]. Still, despite it brilliance, *A canticle for Leibowitz* was as deterministic as the view of any technophile, it was just a determinism leading in the opposite direction. The lesson of our survival during the nuclear madness of the Cold War wasn't that we were fated to survive, but that there was no determined outcome that we would destroy ourselves either.

With the decline in the risk of nuclear war a new progressive technological narrative was able to come into view. This new version centered on the liberating potential of computers and communications networks and was created in no small measure by the refugees from failed Utopias, communes of disillusioned postwar youth who wanted to "get back to the land" and instead discovered a new found appreciation for the power of technology [28].

It is this version of progressive technological determinism that has recently come under increased scrutiny. The charge here is that there might be reasons to be uncertain as to the continuation of technological advancement over the longue durée, beyond the obvious one of self-destruction, and that a laissez-faire attitude to technological development, as with anything else, is as likely to bring outcomes we would not upon reflection want as ones we hope for.

We may tend to assume that our technological advancement will go on forever as long as a global catastrophe does not occur. Yet the silence of a universe fertile for life might give us other reasons for pause [29]. As Lee Billings has pointed out, a non-catastrophic inference from the fact that the effects of other advanced civilizations have not been observed is that we are much closer to some technological peak than we think. The kinds of exponential growth we experienced since the industrial revolution might be a short lived period and a historical aberration [26].

Some have questioned whether the very pace of technological takeoff that helped give rise to middle class society hasn't begun to slow now that the "low hanging fruit" of industrialization have been picked [30]. The future which we imagined with the optimistic certainty seen in the gleaming technological visions of the middle of the

20th century has become increasingly opaque. We have chosen less to reach outward deep into space and time in civilization transforming projects than to turn our gaze inward to measure and monitor ourselves [31].

Perhaps, Laplace's demon wasn't, as it was thought, killed by advances in scientific understanding towards entropy, irreversibility, emergent properties, chaos or complexity [32], but reappeared as efforts at the omniscience of "big data" and the rule of algorithms. Rather than using our increased computational prowess and improved artificial intelligence to build a human future extending outward before us in time and space we have used it to enable a society of mass surveillance that seeks Laplacian omniscience by sucking in and compiling all the minutiae of the present [33], the world's fastest supercomputers used, not to solve the problems of our long term survivability, but to slice time into such small sections they are not even perceivable by human beings [34].

There are also growing doubts over whether technological advancement by itself continues to serve as the foundation for middle class society. Technological development and general prosperity have seemed to have become de-linked, and the budding revolution in artificial intelligence and robotics threatens to pressure what is left of this linkage between improved technology and the support of middle class societies to the breaking point [35].

Most importantly, many are asking fundamental questions about not so much what it means to be human *as what we want being human to mean* in light of emerging technologies. These fundamental questions regarding things such as what the role of memory is to our sense of meaning [36], or privacy [37], or work, [38], or relationships [39], or even war [20], are being asked not only because technology is moving intimately closer to our humanity, but because we really do have choices regarding how this particular phase of technological development will unfold in a way we have not before. It is not the mind-blowing technological powers we continue to produce that count so much as whether we use them to create and support the kind of societies we want.[1]

In some very real sense we may have more room for choice in regards to technology which prior ages have lacked. Industrialization may have been effectively irresistible once it started to gain momentum. Almost overnight in historical terms an enormous number of human beings were pulled off the bottom rung of Maslow's hierarchy of needs where they had struggled since the beginning of history. The best option really was to barrel down on the premise even though technology appeared to be leading to some quite frightening outcomes. Given our already high state of development this need not continue to be the case [40].

All this by a very circuitous route brings me back to the topic of Utopia.
Utopia in its ancient sense disappeared when technological evolution lead us to think that history had a direction, when we needed to and could rely on the advance of our technological powers to free us from the grip of necessity. We are now at a stage where the outcome of simply letting the development of technology continue without our shaping it to better answer our challenges and fit our values is no longer viable.

---

[1] This, of course, is the reverse of Kevin Kelly's focus on what technology wants.

We need something like the idea of Utopia for this shaping. We need it as both a prototype and moral template where many of the problems we currently face are resolved. For none of the current institutions we possess are likely to up to the demographic, environmental or social challenges we face. Our political and economic institutions are in some cases *centuries* old. Yet, public caution when it comes to radical change has a great deal of wisdom in it. We don't know what solutions will work and what they will look like in the real world, or if the cure will end up being worse than the disease. Indeed, the very non-deterministic, non-linear nature of human affairs ensures that *we cannot know* the answers to these questions beforehand.

What we need is ways to test our ideas and examples of solutions that people can actually see then applying what has been shown to work to their own society. Almost all of these experiments will *fail*. Yet their failure is almost the point. Small scale utopian experiments can take the risks of radically innovating while the larger society can use these innovations to engage in what Popper called "piecemeal social engineering" [15] a much less risky endeavor.

Utopians in the 19th century tried this and there are stirrings that some would like to try it again. Today, the right has latched onto this need for social innovation [41] The problem here is that their utopian experiments represent a pretty narrow ideological spectrum. For us to gain much of anything from utopian experimentation we will need such experiments to be much broader.

In some ways we already have such experimentation as a consequence of our fractured political world but we also need more radical experiments. As in natural ecosystems, we could benefit from more even greater diversity in how technology is used and modernity expressed, diversity that would not only give us wide expression for what being human means, but offer us resilience should technological civilization face some existential crisis.[2]

On the purely intellectual level, an image of the perfect society provides us with a moral compass and a tool of comparison to judge the flaws of our own society. In trying to imagine what a perfect society might look like we can become aware of the flaws of our own social systems, conscious of what it is we need to fix or reform. Without some idea of our intended destination we become the plaything of events and risk drifting into shoals we might have otherwise avoided.

The most famous utopian of them all, Thomas More, understood this. His Utopia was in no sense meant as a blueprint for a perfect society, but a means to clarify the flaws in his own [5]. We need to recover our sense of comfort and ease thinking in Utopian terms and rediscover the usefulness of imagining outcomes that are likely unreachable. The "perfect is the enemy of the good" only when our image of the perfect prevents the good from being pursued.

Utopians have always been mental travelers and chrononauts. Plato went backward rather than forward in time to find his ideal city, both Thomas More and Francis Bacon found either a counter or an alternative to their own societies across the wide waves of the sea. Edward Bellamy pulled a Rip Van Winkle and found his Utopia

---

[2]For the role of diversity in system resilience see: Norberg [42]. In terms of the Utopian Tradition see Walzer [43].

in the future after a 100 year nap. The problem now is that utopians have seemingly nowhere, except perhaps into the cold vastness of space to go. Elon Musk, amongothers, is attempting to revive our interest in settling other planets, but even should he succeed the vast majority of us will remain here in this world with all its flaws and injustice.

If Plato, surrounded by war, was able to imagine a way of organizing society that would make war more rare, if Francis Bacon was able to imagine a world where the mass of people were no longer condemned to a life of sickness and poverty in a world that had always had more than its share of both, if abolitionists utopians were able to create small worlds of racial equality amidst societies where fellow human beings were sold and treated worse than cattle, then we are certainly capable of breaking out of the illusion of inevitability which any long lasting social arrangement brings.

No human society will ever truly be a Utopia, but, as Oscar Wilde knew the Utopian imagination has continually expanded our moral horizon. Recovering it might help restore our sense of being creatures embedded in *time* where our agency is directed in the present towards a future whose shape in not yet determined. The future is neither completely ours to shape nor something we are subject to without room for maneuver. For, continuing to think that our world cannot be made to better conform to our ideals is one of the surest ways to insure that what lies in our future is the farthest thing from Utopia. And so, if I were to answer the question that inspired this essay "how should humanity steer the future?" directly, I would say that the question has no definitive and final answer but begins with the rediscovery that it is us with our hands behind the wheel.

References

1. Pinker, S.: The blank Slate: The Modern Denial of Human Nature. Viking, New York (2002)
2. Miller, L.: Fresh Hell: What's behind the boom in dystopian fiction for young readers? *The New Yorker*, June 14 (2010)
3. Cornford, F.M.:The Republic of Plato. Oxford University Press, London (1945)
4. Schaer, R.: Utopia: the Search for the Ideal Society in the Western World. The New York Public Library, New York (2000)
5. Bénéton, P., Paul J. Archambault.:*The kingdom suffereth violence: the Machiavelli/Erasmus/More correspondence and other unpublished documents*. South Bend, Ind. (St. Augustine's Press, 2012)
6. McKnight, S.A.: The Religious Foundations of Francis Bacon's Thought. Columbia, Mo. (University of Missouri Press, 2006)
7. Schaer, R.: Utopia: The Search for the Ideal Society in the Western World. The New York Public Library, New York (2000)
8. Clark, C.: Letters from an American utopia: the Stetson Family and the Northampton Association, 1843–1847. (University of Massachusetts Press, Amherst 2004)
9. Tarnoff , B.: Magical Thinking . *Lapham's Quarterly*, September 23, 2011. See also Friedrich Engels *Socialism: Utopian and Scientific* Engles, Friedrich. Socialism: Utopian and Scientific. Project Gutenberg. http://www.gutenberg.org/ebooks/39257 (accessed April 15, 2014)
10. Pagel, M.D.: Wired for Culture: The Natural History of Human Cooperation. Allen Lane, London (2012)

11. Smolin, L.: Time Reborn: From the Crisis of Physics to the Future of the Universe. Allen Lane, London (2013)
12. Augustine/, Dods, M.:The city of God. Modern Library ed. New York: Modern Library (1993)
13. Morris, I.: Why the West rules- for now: the patterns of history, and what they reveal about the future. Farrar, Straus and Giroux, New York (2010)
14. Marqui, L., Pierre, S.: Philosophical Essay on Probabilities, p. 4. Hardpress Ltd, S.l (2013)
15. Popper, K.R.:The Open Society and Its Enemies, 5th ed. Princeton, N.J.: Princeton University Press (1966)
16. Levin, M.R.: Ameritopia: the Unmaking of America. Threshold Editions, New York (2012)
17. Armstrong, K.: The Case for God. Knopf, New York (2009)
18. Ridley, M.:The Rational Optimist: How Prosperity Evolves. New York: Harper (2010)
19. Searle, R.: Silicon Secessionists. Utopia or Dystopia. http://utopiaordystopia.com/2014/01/12/silicon-secessionists/ (accessed April 15, 2014)
20. Coker, C.: Warrior geeks: how 21st-century technology is changing the way we fight and think about war. Columbia University Press, New York (2013)
21. Kelly, K.: What Technology Wants. Viking, New York (2010)
22. Diamandis, P.H., Steven K.:Abundance: The Future Is Better Than You Think
23. Ray Kurzweil Explains the Coming Singularity | Big Think. Big Think. http://bigthink.com/videos/ray-kurzweil-explains-the-coming-singularity (accessed April 15, 2014)
24. Morozov, E.: To Save Everything, Click Here: the Folly of Technological Solutionism
25. Greene, J.D.: Moral Tribes: Emotion, Reason, and the Gap Between us and them. New York. Penguin Books (2014)
26. Billings, L.: Five Billion Years of Solitude: The Search for Life Among the Stars S.l. New York Penguin Books (2014)
27. Miller, W.M.. A Canticle for Leibowitz. Bantam trade pbk. ed. New York: Bantam Books (1997)
28. Turner, F.: From Counterculture to Cyberculture: Stewart Brand, the Whole Earth Network, and the rise of digital utopianism. University of Chicago Press, Chicago (2006)
29. Sasselov, D.D.: The Life of Super-Earths: how the hunt for alien worlds and artificial cells will revolutionize life on our planet. Basic Books, New York (2012)
30. Cowen, T.: The great stagnation: how America ate all the low-hanging fruit of modern history, got sick, and will (eventually) feel better. Dutton, New York (2011)
31. Rushkoff, D.: Present Shock: When Everything Happens Now : Penguin Books, 2013. See also: Billings, Lee. *Five billion years of solitude: the search for life among the stars* S.l.: Penguin Books (2014)
32. Ulanowicz, R.E.: Growth and Development: Ecosystems Phenomenology. Springer, New York (1986)
33. Bamford, J.: The NSA is Building The World's Biggest Spy Center: Watch what you say. *Wired Magazine*, March 15 (2012)
34. Patterson, S.: The Quants: How a Small Band of Math Wizards Took Over Wall St. and Nearly Destroyed it. New York: Crown, 2010. See also: Lewis, Michael. *Flash boys: a Wall Street revolt.* : W.W. Norton and Company Inc, (2014)
35. Lanier, J.: Who Owns The Future? Simon and Schuster, New York (2013)
36. Garber, M.: This Email Will Self-Destruct After You Read It'. The Atlantic. http://www.theatlantic.com/technology/archive/2014/04/this-email-will-self-destruct-after-you-read-it/360470/ (accessed April 15, 2014)
37. Scharr, J.: Vienna Teng Sings about Surveillance in 'Hymn of Acxiom'. Tom's Guide: Tech For Real Life. http://www.tomsguide.com/us/vienna-teng-hymn-of-acxiom,news-17663.html (accessed April 15, 2014)
38. Lanier, J.: Who Owns The Future? Simon and Schuster, New York (2013)
39. *Her*. Directed by Spike Jonze. S.l.: Warner Home Entertainment (2014)
40. Searle, R.: Privacy Strikes Back, Dave Eggers' The Circle and a Response to David Brin. Utopia or Dystopia. http://utopiaordystopia.com/2014/03/09/privacy-strikes-back-dave-eggers-the-circle-and-a-response-to-david-brin/ (accessed April 15, 2014)

41. Thiel , P.: The Education of a Libertarian. Cato Unbound. http://www.cato-unbound.org/2009/
    04/13/peter-thiel/education-libertarian (accessed April 16, 2014). See also Paul Romer: New
    cities. More Choices.Better rules.. Charter Cities. http://urbanizationproject.org/blog/charter-
    cities (accessed April 16, 2014)
42. Norberg, Jon. Diversity and Resilience of Social-Ecological Systems coauthored by Nobel
    Laureate Elinor Ostrom In *Complexity theory for a sustainable future*. New York: Columbia
    University Press, 2008
43. Walzer, Richard. *Al-Farabi on the perfect state: Abudila: a revised text with introduction,
    translation, and commentary*. Oxford Oxfordshire (Clarendon Press 1985)

Chapter 8
Enlightenment Is Not for the Buddha Alone

Tejinder Singh

Abstract The second law of thermodynamics provides the universe with an arrow of time. Living organisms are metastable states which, through the process of aging, are also subject to the second law. At the top of the living chain is humanity, with the human mind and the creative thought process possessing a tremendous ability to alter its environment. However, the mind is also an inefficient storehouse of redundant, repetitive and unproductive thinking. Combined with the mind's acute ability to remember the past, and think about the future, such unproductive thinking can become a source of harmful negative emotions such as anger, hatred, worry, anxiety and fear, amongst others. It is possible to overcome such unpleasant consequences resulting from the ever-thinking mind, by realizing that there is a underlying thought-less state—Consciousness. An individual who operates from the state of a conscious I, then lives in the Here and Now, and is happier, and at peace with oneself, and more likely to contribute constructively and compassionately to the task at hand. It is this state that humanity should collectively strive to steer towards. If this can be achieved, even to a partial degree, it will become easier for humankind to address and resolve the practical threats and challenges we face on our planet today.

The universe is a most extraordinary place. It exists of its own, in an objective manner, driven by the laws of physics, some which we understand, and some which we are yet to discover. And yet, in spite of all the objectivity we attach to it, we cannot dissociate the objective universe entirely from the human mind which tries to comprehend it. Who is to say that the mind is definitely possessed with the capacity to unravel all physical laws, and that eventually we will get there. Who is to say with certainty that one day we will know why the universe is there in the first place? Be that as it may, it is our paramount duty to protect and nurture the mind, and make it more and more capable of seeking out those truths. For this, a collective improvement on part of all of humanity is called for, involving not just scientists and mathematicians. For only by simultaneously addressing the more practical concerns that humanity faces, can we make the planet a more conducive place for advancing our investigations of the laws of nature.

T. Singh (✉)
Tata Institute of Fundamental Research, Homi Bhabha Road, Mumbai 400005, India
e-mail: tpsingh@tifr.res.in

© Springer International Publishing Switzerland 2016
A. Aguirre et al. (eds.), *How Should Humanity Steer the Future?*,
The Frontiers Collection, DOI 10.1007/978-3-319-20717-9_8

8.1 Time and the Inanimate World

Physics teaches us that matter and fields live in space and evolve with time. At least in the classical Newtonian world. [And also in the semiclassical world, where matter fields are quantized and the gravitational background is classical.] And that is how the human mind perceives it too—a past that has happened, an instantaneous present, and a future that is yet to come. In the microscopic world, evolution is time reversible; but in the macroscopic world the second law of thermodynamics is inexorably at work, and there is an arrow of time.

Physics also cautions us that when quantum gravity comes into play, there may be no time and no space in the conventional sense, let alone an arrow of time. This perhaps happens on the Planck length scale and on Planck densities, and perhaps in some other physical circumstances too. For instance, who is to say that one can with certainty talk of causality when the collapse of the quantum wave-function at one point in space seems to have an instantaneous influence on a space-like separated event? But the human mind seems incapable of tangibly grasping any such timelessness—or at least so we commonly believe: more on this shortly.

Where does such time irreversibility in the macro-world come from, when the micro-world is time reversible? We believe we can only encode this in the initial conditions with which the universe began. For reasons that we do not quite understand, the initial entropy of the universe was far, far less compared to what it could have been, and then the unavoidable expansion in phase space causes the entropy to increase, providing us with the observed arrow of time.

This law of ever increasing entropy holds with such infallible consistency that we have convincingly applied it to every conceivable circumstance, including the world of living things.

8.2 Time, Living Beings, and the Human Mind

In simplistic terms, we might think of a living organism as a material state in metastable equilibrium, which maintains itself, for a certain length of time, in a state of low entropy, by consuming low entropy nutrients [food] and by subsequently increasing the entropy of the environment. It is safe to assume though that the metastable state is not forever—it is subject to aging. This aging in itself is a clear signature that within a living being too, the second law is at work, and its entropy slowly increases. The end result of the metastable state is inevitable death: the second law wins over; there is no living organism that lives forever. But living beings have found a clever trick to ephemerally beat around the second law—reproduction! Every time a new organism is born of the parents, it starts afresh in a low entropy state, and the cycle repeats itself over again, and on it goes. It is clear then, that in this birth-death-birth-... cycle the arrow of time is very much in play.

Undoubtedly this scenario applies to human beings too, but we know that it gets more intricate as we make a transition from lower level living forms to higher ones, including mammals, and humans. Growth and aging in lower living forms presumably follows an automatic pattern, devoid of free will. By the time the ladder of life reaches mammals and in particular humans, something entirely novel comes into play: a highly evolved brain, and with it, a thinking mind. Mind that perceives time, mind that has a memory of the past, a perception of the present, and an anticipation of the future. Mind which with its great capacity for creativity and innovation, has changed the face of the planet. Mind that understands the universe, discovers and formulates physical and mathematical laws of nature, and builds technologies and civilizations. Mind that can be compassionate and spiritual.

And yet, mind that is also a garbage dump of thoughts! Spanned across humanity, most minds, at most of the times, are repeatedly thinking the same thoughts over and again, thoughts that are not only of no consequence, but sometimes harmful to the well-being of the body and the being. Emotions that are irrational and not founded on the factual situation of the immediate present: traumatic memories, unhappiness, anxiety, apprehension, uncertainty, depression, fear, including fear of death. Fear of other human beings, often unfounded. Undoubtedly it is the case that sometimes such thoughts and emotions are precipitated by real life events, including natural calamities. But more often than not, the mind worries, and worries without reason.

Mind that is evil. Mind that plans warfare. Mind that plans and executes killing of human beings. Millions of humans were wilfully killed by other humans in deliberate acts of violence in the previous century alone. No other species destroys its own on such vast magnitude. Mind that is greedy. Mind that is bitter and hates. Mind that is often unsatisfied, no matter how much has been acquired, gained and possessed. We cannot deny that a good fraction of humanity lives in this mental state, to some degree at least, while it is undeniably also true that a good fraction of humanity is genuinely suffering, from poverty, disease, misrule, and other unpleasant causes.

What is the origin of the thought clutter in the mind, which makes it so inefficient, and the origin of the hurtful, unpleasant thinking? The origin lies in the state of biological evolution humanity currently is in. Long, long ago the characteristic of intelligent thinking began to evolve, which gave us the ability to innovate, and master the environment. But along with it came the 'thermal noise' of thought—as if nature were trying to make a perfect thought machine, but only succeeding in making an inefficient one, at least as of now. One could be certain that in the very long run the process of evolution is progressing towards converting human minds into more intelligent and efficient ones, where the noise will progressively reduce. One could envisage the arrival of an evolutionary stage where the mind thinks only that which is essential. But for now our civilization is stuck with this thought noise, and it is hurting us bad. For when nature innovated the thinking mind, it also possessed it with a memory of the past, and a sense of the future yet to come. Change, growth, aging and the second law are very much at play when it comes to the thinking mind.

This combination of thought noise, and the ability to perceive the flow of time, is often a recipe for disaster, and the source of the ills alluded to above. While on the one hand the mind and body are truly speaking existent only in the present moment [the here and now] the noise of irrational thoughts relentlessly and continuously forces the mind into thinking and worrying about the past and the future [the there and then]. This flight of the mind to the there and then, when it should be here and now, is predominantly the birthplace of unhappiness, anger, hatred, greed, sorrow, depression, anxiety and fear. True, these emotions could even be precipitated by some mishap in the here and now. But we know that these are rarities; on most occasions the here and now is in reality peaceful, but the worrying mind turns it into an unhappy present moment. In short, the mind is unhappy because it is afflicted with thought noise, and because it comprehends the flow of time.

We take this affliction of the uncontrolled mind for granted; we do not even think of it as a disease, because everyone has it to some degree or the other. Now of course not all of humanity is always unhappy! And sometimes there are genuine causes of unhappiness. That is not what we are referring to. We are talking about the noisy mind being a source of unnecessary and unfounded worry, which affects most people sometime or the other. One instance is the imagined fear of death, which fear lurks even when death is not evidently near—this is a prime example of thought noise projecting itself into the future to conjure something primordially perceived as very unpleasant. Even if a mind trained in physics might know that the second law will ultimately take the body to death, that knowledge is not necessarily sufficient to mitigate the fear. What to talk of the untrained mind then!

How does humanity overcome this affliction? Are we to wait for biological evolution to take (wo)mankind to the state where the mind is efficient and free of noise? No. A handful of Enlightened seekers over the past few millenia have shown us a way out. But while they have told us that everyone can in principle seek and achieve what the seekers have done, clearly that has not happened. Most of humanity does not consist of Enlightened seekers; most humans are perhaps not even aware that there is an escape possible from this unpleasant noise, many are perhaps not even interested! The conditioned mind, whose master is thought, then often uses thought for its design and intent, often evil, often with the express desire for power and control over others. At the end of the day, this only creates more and more global unhappiness and imbalance in the environment, in spite of all the progress and comfort that science and technology have given to us. We have all in our lives come across someone or the other who appears extremely calm, quiet and content, happy and compassionate. In whom the thought noise and worry does not seem to be there. Perhaps he/she was a spiritual teacher. A noteworthy modern day example is the Zen Buddhist monk Thich Nhat Hanh. One look at his face, and one listening session with him, is enough to convey the deep inner peace that prevails in him. It is practical, important and significant to ask why everyone is not at peace like him? Does one have to be a renunciating monk with extraordinary mental powers, to free the mind of its dreadful noise and the consequent troubles? No. Certainly not. With some help, or on our own, we can all get there, to some degree at least. And that is where humanity should steer to, so that we have more and more human beings who are happier, and at peace with themselves.

8.3 The Timeless Consciousness

I go and lie down in a quiet, dimly lit room, with no one else around, perhaps at night, when it is still, and I close my eyes. I become aware of my body, and of my senses. I become acutely aware of my surroundings, alert to any small noise in the surrounding. I become aware of my regular rhythmic breathing, and I focus on my breath. Amazingly, when one is focused on one's breath, one cannot think. This takes attention away from the mind, and thinking slows down. With some practice and training, thinking can be stopped nearly altogether, and the incessant thought clutter almost won over. The mind goes blank, literally. I am no longer worrying of the past, nor imagining the future. I am simply in the present moment, and at peace.

At the same time, something else extraordinary happens too, which can hardly be described easily in words. One becomes aware of oneself, in a manner which goes beyond thoughts and mind, for the thinking mind has been stilled already. One becomes aware of oneself as a whole, and one can watch oneself as if from 'outside', and one can watch over the mind. If a thought comes, one can watch it come and go. This state of self-awareness one may call Consciousness. It is the ultimate identification with the I, with the Self. Once such awareness has been grasped, and dissociated from the mind, one could stay identified with it for as long as one is alive. Or if, having grasped it, one loses it, one can come back to it, for one knows what one is looking for. This state of Consciousness is the timeless 'I'. And it is spaceless too. For when the mind has been stilled, there could be no memory of the past, and no anticipation of the future. There is only the Consciousness, and only the immediate present in which the Consciousness resides. For the Consciousness, the world is timeless, every moment when it comes is the present moment. This is another instance, apart perhaps from quantum gravity and acausal EPR correlations, where one encounters timelessness, though we are not at all suggesting that these instances might be related in some way. We have to wait until we have a scientific understanding and model of Consciousness. But we could be sure that Consciousness is some timeless feature of a living system such as a human being, which is undoubtedly a property or state of matter, but it goes beyond body, brain, mind and thought, in a way that we do not understand today.

That the conscious I is timeless one could be certain of. For, everything else changes—the body changes and grows older, thoughts and emotions are in a continuous state of flux. The cells of the body regularly die and new cells are born. And yet there is one constant in a human being: the conscious I. It stays with us throughout life—I am the same I irrespective of how young or old I am. It is to be contrasted with the time-bound I, which is associated with the mind, and which changes continuously. The conscious I does not age during life—it is obviously the same sense of I always. Is this in defiance of the second law? Or is it an equilibrium state—the timeless Consciousness? Intriguing, but we do not know. We could not know, until we mathematically and scientifically understand Consciousness.

The realization and achievement of this state of Consciousness, which is timeless, and where the mind is still and free of thought, we may define as poor man's

Enlightenment. It is a beautiful peaceful state, for if the mind is not thinking, there is no worry, no anxiety, no fear. There is only a powerful deep-rooted identification with the present moment, which moment is infinitely simpler than the gigantic burden of one's past, and simpler than the uncertainty of one's future. If we can identify ourselves with this conscious timeless I, we become masters of our minds, we start to control our thoughts, instead of being controlled by our thoughts. A thought may come, but Consciousness decides whether to follow it, rather than being enslaved by the relentless running and repetition of the thought. Psychological suffering and psychological pain goes, for they are part of the active mind. Bodily pain, if any, would remain, but even that would feel less so, because the psychological component will have been removed. There is no denying that one who lives in the present moment, by the Conscious timeless I, and not by the noisy thought-pressed mind, gets to control the mind, and is a happier, less worried person. It is like becoming someone who watches oneself from the outside, deliberately controlling one's thoughts and actions, rather than letting them happen of their own accord.

It is all very well, one might say, to achieve this beautiful Enlightened state, lying down meditative with closed eyes in one's bedroom. Of course we all very well know that is not real life. Welcome to the world of people, interactions, confrontations, disagreements, dangers. What happens now? Two principles that were learnt in the bedroom with closed eyes must be adhered to. Operate and act from the timeless Conscious I, and stay deeply rooted in the present moment. Thoughts will come and go, but the superior Conscious I, which watches over the mind, lets the thoughts be, picking out only those which one wishes to, or those which are of consequence to one's actions, and to the advancement of one's intellect. The creative process germinates from the Conscious I. One thinks of the future to the extent that one plans for it constructively, but one does not worry about the future. Deliberate decisions are taken by the Conscious I, not by the wandering conditioned mind. With practice and training, this is possible for us all common people. The Conscious I, which is naturally at peace with oneself, will see no reason to harm others, or to be beleaguered by negative emotions. The rooting in the present moment simplifies life extraordinarily. It takes away the traumas and bitterness of the past. It takes away anxiety. It takes away the fear of death. It makes one compassionate towards other humans. The rooting in the present moment also brings an extraordinary awareness and alertness of the surroundings, in which one sees beauty, and which act as a source of joy, howsoever mundane and monotone the surroundings might be. It makes one a happier person. If everyone of us were to be like this, would the world not be a happier place to be in?

Easier said than done. Too idealistic. But at least we could try. Right now there is no global effort in this direction, no mass movement. Only a few spiritual teachers, and their disciples, almost as if working in isolation from the rest of the society at large. And clearly we are not talking of established organized religions here, in the conventional sense, where of course there is enormous following and appeal. Whether that of its own has helped humanity, the reader can judge for himself/herself! It was always the express purpose of all religions, when they were born, to communicate the deep relevance of Enlightenment, of the Conscious I, and the importance of being in the here and now. Unfortunately, with time and history, these important messages have to a large degree been lost and buried, and replaced by something else…often ritualistic and irrelevant, the most glaring misfortune being conflict amongst religions, something furthest from the original message.

8.4 Towards a Conscious Humanity

Humankind's primary concerns are eradication of poverty and disease, imparting education, development, controlling damage from natural calamities and climate change, search for alternate energy resources, good governance and peacekeeping amongst nations, scientific advancement and technological innovation, looking into the possibly of efficiently migrating into outer space and to habitable exoplanets, and other conventional modes of economic progress which seek to create better societies.

Given that these above are the priorities, is talk of Enlightenment impractical mumbo-jumbo? The answer to this question should be sought in another question. While reasonable success has been achieved in pursuing many of the above concerns, and in particular in science and technology, has this success made us all into happy individuals? The answer clearly is: only partly so, very partly so. What is the key reason for achieving only partial success? It is certainly not for want of resources, of which there were aplenty to begin with. Nor is it for want of innovation, where we have seen staggering strides. The key reason is the uncontrolled human mind! An uncontrolled mind which often applies itself destructively to the individual and to the society. And acts as an obstruction to the better half of the mind which is working hard to address the above concerns. We have with resignation accepted this uncontrolled and confused mind as a curse of nature. However we have seen above that when an individual operates from the conscious I, instead of operating from the level of the mind, and lives in the here and now, the 'uncontrol' and 'confusion' vanish, and are substituted by peace and compassionate action. Hence progress towards individual Enlightenment is not mumbo-jumbo talk, but a progress which will in a very healthy and productive manner enhance our ability to address the above primary concerns. Minds that are free of wandering thoughts are minds that are more creative, more constructive. Thus we are not talking of people retiring to monasteries en masse; we are talking of people become enlightened monks as they go about their daily lives taking care of their homes and professions.

What is the recipe for creating a globally Enlightened humanity?! Clearly there are no easy answers. A social revolution would be required. And that must be preceded by political will to permit such a revolution in the first place—an act of contradiction by itself, for it calls for using power to surrender power!

Assuming that those in governance are enlightened enough to allow progress in this direction (and that is an enormous assumption), one can think of baby steps to implement. Schools and primary education would be the first place to start from. It is not far-fetched to imagine that one can talk to young teenaged children about mind, thinking, meditation, and deep breathing. Insightful text books written by contemporary spiritual teachers and introduced into the school curriculum would be of great help. A school text book on Enlightenment on par with text books on physics, biology and mathematics? It definitely sounds fanciful and perhaps even ridiculous. But we are talking of nothing short of a revolution here, if we are to steer all of humanity towards this so-called enlightened direction, and from the viewpoint of the establishment, a revolution often does appear fanciful, ridiculous and unacceptable. We have little choice but to start with school. The subject would not be so much about learning and conceptualisation, but about practice and training, day after day, year after year. And won't we be making them happier teenagers in the process? We certainly would be—currently they are one of the most confused and unsettled section of our society. Who would be the teacher? How can a teacher who is himself or herself not enlightened, talk about and train others in Enlightenment?! There is no easy answer. Perhaps the teacher who teaches value education in school must learn this 'subject' as he or she goes along, and perhaps visit and learn under a spiritual teacher.

The next group to be addressed, perhaps the prime group—young adults—are students at college and university. Apart from compulsorily introducing such teaching in regular curriculum, there would be ample scope through voluntary extra-curricular initiatives. Meditation, consciousness, and mastering one's thinking, are definitely concepts that young minds can be taught and trained in.

Then comes the office and the work place, both in the government sector, as well as the private sector. Adults with homes, families, and children. Spirituality classes should perhaps be made mandatory at the work place. Undoubtedly they will improve the working atmosphere and inter-personal professional relationships.

That still leaves a large fraction of the population—people who are self-employed, and people who stay at home. In reaching out to them, an enlightened media can play a far-reaching pivotal role, by proactively creating awareness. Smooth access to such enlightenment education can be provided via the internet, which can also provide online resources for use at the work place and in colleges and universities.

In passing, it is amusing to note that spirituality is widely perceived as the domain of the old, for whom the mind somehow has reached a stage where search for Enlightenment comes far more inevitably than it does for youth and for the middle-aged. One could be certain though that the need for such education is far more pressing for the youth and the middle-aged, than it is for the old.

Needless to add, those who are poor and hungry, or are beset by natural calamity, have more urgent needs to be attended to, as compared to Enlightenment.

I am acutely aware of how fanciful and improbable such attempts at implementing global Enlightenment might sound. But what other way can there be, if one is to go beyond only a handful of individuals seeking out on their own, and converting such seeking into a mass movement which will help humanity at large.

In attempting to initiate such a mass movement, humankind would only be helping the course of natural biological evolution. As we noted earlier, the human mind will most certainly evolve towards a stage where inefficient thought noise will reduce, producing minds which think less and are more efficient. Very likely, this will bring Consciousness to the fore, evolving us to a race where the I dictates thinking, and not the other way round. We have become conscious enough to realize this, so what better than to do unto ourselves right now what nature is going to do us in the long run. And who knows, unless we this unto ourselves, the evil mind could self-destruct long before nature reaches us to that Enlightened state!

We must also not forget that scientific investigations of the thought process, mind, and Consciousness are also likely to lead us into realizing the existence of this Enlightened state. These investigations will most probably lie at the fascinating interface of neurobiology, biochemistry, quantum theory, thermodynamics, and condensed matter physics. A scientifically sound mathematical model of Consciousness will compel us to accept the existence of the timeless conscious I, which rules over the thinking mind. It is then not pure speculation to suggest that the fields of medicine and psychiatry will themselves encourage and educate individuals to seek out the Enlightened state and become more peaceful, happier beings. One cannot think of a more beneficial and rewarding confluence of science and spirituality.

8.5 Why Is It so Difficult to Make This Work?

All I really need to know about how to live and what to do and how to be I learned in the kindergarten. Wisdom was not at the top of the graduate-school mountain, but there in the sandpile at Sunday School. These are the things I learned. Share everything. Play fair. Don't hit people. Put things back where you found them. Clean up your own mess. Don't take things that aren't yours. Say you're sorry when you hurt somebody. Wash your hands before you eat. Flush. Warm cookies and cold milk are good for you. Live a balanced life—learn some and think some and draw and paint and sing and dance and play and work every day some. Take a nap every afternoon. When you go out into the world, watch out for traffic, hold hands, and stick together. Be aware of wonder. Remember the little seed in the Styrofoam cup: The roots go down and the plant goes up and nobody really knows how or why, but we are all like that. Goldfish and hamsters and white mice and even the little seed in the Styrofoam cup—they all die. So do we. And then remember the Dick-and-Jane books and the first word you learned—the biggest word of all—LOOK. Everything you need to know is in there somewhere. The Golden Rule and love and basic sanitation. Ecology and politics and equality and sane living. Take any of those items and extrapolate it into sophisticated adult terms and apply it to your family life or work or your government or your world and it holds true and clear and firm. Think what a better world it would be if we all—the whole world—had cookies and milk about three o'clock every afternoon and then lay down with our blankies for a nap. Or if all governments had as a basic policy to always put things back where they found them and to clean up their own mess. And it is still true—no matter how old you are—when you go out into the world, it is best to hold hands and stick together.

Robert Fulgham

We are all born enlightened, in the sense of being rooted in the Here and Now. That is evident from the happy playful infant who engages only in its immediate surroundings. In our early years also we are enlightened. What changes afterwards?

Where does the enlightenment vanish? While this is an extremely complex and difficult question, a simplistic answer is the growth of intelligence as the child grows. With intelligence comes thinking, with thinking come understanding and academic growth, and unfortunately also anxiety, worry and fear. The biggest compounding factor is inter-personal rivalry and conflict. Classroom performance and success in education becomes competitive and judgemental; coming first in classroom exams is a matter of great pride and achievement; and performing badly brings negative self-image. [I speak from an Indian perspective, having observed what happens in my country.] Students are judged by how well they do in subjects as diverse as history, civics, geography, physics, chemistry, biology, mathematics, languages, computers, environmental science, amongst others. He or she who does best in them all is adjudged a hero, no matter if this was rote learning, learning under pressure, learning not for joy and pleasure, but for doing well in exams. The young mind has been bludgeoned, creativity often stifled, and all except the brightest and the most creative, who seem immune to the torture, succumb to the so called exam pressure. The Great Test comes at the end of ten years of school, when the so called nerve-wracking Board exams are held, and the marks you obtain there are stamped on you for the rest of your life. If you did not score above 90 %, you are a dud. Ridiculous extremes are reached when parents clamber over ladders and school walls to pass on cheat sheets to there wards in the examination halls, who merrily copy from them with overt assistance from the invigilators. What are we trying to prove here? What have we achieved? We have destroyed the enlightened, innocent child.

If only we did it a little differently. If only we kept up with Fulgham's teachings beyond kindergarten. We need to teach life skills, build a high emotional IQ, help them learn to cope with life situations. The real world in which we adults go about our daily personal business has so little to do with the 'history, civics, geography, physics, chemistry, biology, mathematics, languages, computers, environmental science' we learnt in school. True, these subjects are a part of our essential knowledge base, but why did no one ever teach us life skills in school? And this would include teaching us about being in the here and now, about the conditioned mind, and about the Conscious I. It will not take more than some 10 % of the teaching, to be exchanged say for just one other subject, but the personal gain for the student will be more than 90 %. Somebody in school needs to pay attention to what smartphones, social media networks and the internet are doing to our children. They are making zombies out of them, fingers permanently clicking away at the cell phone, oblivious to surroundings, oblivious to nature and outdoors, oblivious to family and dinner table etiquettes!

I believe, what we teach, how we teach, and the intent with which we teach, and what we do not teach, has a lot to do with creating adult troubled minds. We need to change this. The change has absolutely nothing to do with religion, if by religion we mean belief in a named God or God's representative on earth. Religion is a personal choice made at home; to be kept apart from the life-skills taught at school. And we all know the tragedy of religion. Multiplicity of Gods, communal hatred, and organised warfare motivated by religious strife. If only the whole planet had one God and one religion (if at all human beings continue to have a need for God and religion) there would be less mayhem all around. We may call the school reforms spirituality classes or enlightenment studies, but if that name carries overtones associated with one or the other religion, we may simply call them life-skills classes. And life-skill training also ought to majorly include teaching equality of religions, religious tolerance, and the very meaning of religion itself. And such training has nothing to do with East versus West. It is neither Oriental nor Occidental. It is basic common sense, in a manner of speaking!

In asking who will initiate this change and school reform, we have a great chicken and egg problem. Change has to be initiated and approved by governments, but governments are made of the same minds which have gone through the unreformed school. How will they ever appreciate the need for such a change? One can only hope.

Even if there are enlightened individuals in governments, and some surely are there, there are two enormous hurdles, as I see them, which severely hamper attempts at implementing mass enlightenment. One is the concept of Nation State. The other is overpopulation, leading to scarcity of resources, poverty, and disparity in distribution of wealth and resources.

The Nation State is a global curse: we must defend our territories and resources against each other; every nation state is a potential foe against every other nation. Equipartition of wealth and natural resources is unthinkable. Enormous investment is made and wasted in defence, weapons and armed forces; scores of wars are going on at any given time, hundreds are massacred on any given day—man is pitted against man for the sake of defending territory, or forced occupation. See the ridicule of it, from the vantage point of an observer watching our planet intently from outer space: a figment of planetary rock and ocean that we are, a speck in the vastness of the universe, and unmindful of the vastness out there, killing each other. A bunch of power hungry rulers of nation states often scheming diabolical schemes against other nations. Given this, how can even an enlightened ruler train his or her people to live in the here and now, with the peaceful Conscious I, when directly or indirectly the greater concern is to defend our nation from the others?! What good is a nation of enlightened individuals, who are content to not harm others, when their neighbouring countries are just waiting to pounce? Enlightenment at an individual level is still feasible, but as a mass movement it has to begin by nation states seeing eye to eye, and by agreeing to not go to war. Who is to say how that can be achieved? Will an alien invasion unite us together, or will some of us gang up with the aliens, and pit against the rest of us? In the quote above, Fulgham asks governments to be like the children in kindergarten, to clean up their own mess, and to put things back where they found them. How true! Governments and leaders need to see sense, make peace with each other, and initiate life skills training, globally, in schools and colleges. Hopefully there will be more sensible people on our planet then, who will want to reach out to the poorest, who have not even seen much food nor shelter in their lives, let alone schools and education.

Undoubtedly, we have overpopulated the planet. Along with life skills training, schools, colleges, and governments need to educate individuals on the dangers of overpopulating the planet. Communities do not see it as any of their business to talk to families about the number of children they ought to have. Birth is an individual's birthright! The family does not care for the population; the system does not care for the size of the family. This deadlock has to break. In a compassionate way, the state should have a say in the matter—not by way of coercion and decree, but by way of cordial discourse. We take the bursting population of the planet as a given, but it is high time we woke up!

Imagine; if only nation states had a policy not to attack each other (even if they do not agree to share resources or consider a one nation world), and if only they were not overpopulated, it would be much easier then to address concerns of the individual, and make progress towards training individuals to be enlightened.

And while we work towards the bigger goal of enlightenment, it is well for us to remember a few simple tips that go into making us happier; tips that we can share with our children as well. Practice kindness. Express gratitude for what we have. Buy less gadgets! Buy things that create experiences, like a musical instrument. Do not hang out so much on social media. Cut down on checking e-mail! Checking emails every so often creates stress. Realise that time is a precious resource. Let us lose ourselves in some fun activities that we like. Embrace failure. All these tips are nothing but Fulgham's kindergarten lesson all over again!

8.6 Concluding Remarks

We started by observing the paramount significance of the second law of thermodynamics in determining the evolution of the macroscopic universe. For reasons that we do not well understand as of today, there also exist in the universe living organisms—metastable low entropy states which survive by feeding on negative entropy from the environment, while on the whole the second law continues to be obeyed. At the pinnacle of the life chain is the human species, possessed with this extraordinary capacity to think, which has given us the power to dramatically and wilfully change our environment. And yet, because the thinking mind is very inefficient, and lives in time, and remembers the past and anticipates the future, it often self-destructs the species, the worst example being organized warfare. We realize that beyond the disturbed thinking mind there exists the peaceful, timeless conscious I. We advocate that all of humanity should endeavor to realize this conscious I, and operate from that vantage point, which allows us to be deeply rooted in the present moment, control our thoughts, and to be happier, compassionate humans. Only by doing so, can we hope to collectively acquire the competence to overcoming challenges such as poverty and depleting resources, and become a species intelligent enough to successfully execute plans to migrate to outer space.

In 1637, in his search for an infallible truism, Descartes wrote in his 'Discourse on the Method': *je pense, donc je suis*: cogito ergo sum: I think therefore I am. Today we know this to be not true! Even when I do not think, I am. In fact when I do not think, it is then when I truly am. I am then the permanent timeless I, which is always in the here and now. When I think, I am only the wandering time-bound I. It is this timeless I which one seeks, and which when found, binds us with who we truly are, underneath the wandering world of thoughts and emotions. The identification with the timeless I is what makes us truly happy and peaceful, and the state that all of humanity should steer towards.

The author is deeply indebted to Thich Nhat Hanh and Eckhart Tolle for their speeches and writings, from which he has benefited in a very significant way.

Chapter 9
Back to the Future: Crowdsourcing Innovation by Refocusing Science Education

Travis Norsen

9.1 Science and Science Education

Science is a little bit like the USA in the 19th century. There is a core of well-settled territory that is efficiently governed by well-understood and widely-respected laws. Call that "the East". But then there is also "an expanding frontier of ignorance" [1]—in the West—where uncertainty and controversy reign supreme.

But this is not at all the picture of science that students acquire. Textbooks focus on the East—on the well-established conclusions of the past—and largely ignore the contemporary frontier. That is understandable enough since, well, the West can be a confusing and dangerous place, especially for children. But more importantly and less understandably, textbooks systematically suppress the phenomenon of *expansion*— i.e., the fact that every part of what we now call the East was once part of the Western frontier.

Textbooks, that is, tend to present the established conclusions of science as time-less, contextless truths, to be accepted on the basis of formal proofs or, often, on no basis at all besides the sheer fact that they are printed in textbooks. Very little information is given about the methods—the chronological steps—by which the conclusions were arrived at. (And very often, when some such historical explanations are presented, they are highly misleading, biased, and/or totally inaccurate [2].) Science education thus inadvertently tends to make science appear authoritarian and dogmatic.

Addressing that problem is at least part of the motivation behind contemporary educational reform movements, such as those advocating for less lecture and more hands-on, inquiry-based activity [3]. Science teachers' recognition

T. Norsen (✉)
Physics Department, Smith College, Northampton, MA 01063, USA
e-mail: tnorsen@smith.edu

T. Norsen
210 Middle St, Hadley, MA 01060, USA

© Springer International Publishing Switzerland 2016
A. Aguirre et al. (eds.), *How Should Humanity Steer the Future?*,
The Frontiers Collection, DOI 10.1007/978-3-319-20717-9_9

of the problem is also illustrated by the increasing sense of urgency associated with getting students involved in *research* early in their educational careers. Teachers seem to recognize that there is something "fake"—something misleading about the actual character of science—in the textbook-centered classrooms. So, they apparently think, we should get students out of there and let them confront actual puzzles about how best to explain phenomenona, how to interpret unexpected data, how to design appropriate experiments to decide between competing hypotheses, etc. We should, in short, give students a first exposure to the frontier in the West—a first taste of real science.

Getting students involved in research is a good thing. But we will only ever be able to do this for a small minority of exceptional students. Why subject the rest to a "science education" that systematically misrepresents the actual nature of science? Why not bring "real science" into the classroom, from the beginning, so that everyone can learn it, benefit from it, and apply it to the puzzles whose resolutions (or lack thereof) will shape humanity's future?

One possible way of doing this is to radically revise not just *how* we teach, but *what* we teach—in particular to fuse scientific content with scientific method by explicitly teaching the historical discovery process of major scientific conclusions. Such an approach to science education is not a new idea. It has been endorsed by eminent scientists like Albert Einstein, Louis Pasteur, and Ernst Mach [4]. And there have been several major historically-based curriculum development projects [5], most notably "Harvard Project Physics" [6].

The idea of incorporating historical material and perspectives into the science curriculum—so that students focus more on "interesting puzzles and how they were resolved" and less on "truths to be accepted"—seems very appealing. It would clearly make normal science education more research-like and would make a far greater number of students far more able to apply genuinely scientific methods to the puzzles that arise in their own lives and careers. Why, then, has this idea never really caught on, despite its long history? I think the main reason is that it has usually been advocated in the name of "outreach". For example, the major motivation behind "Harvard Project Physics" was to make physics more accessible to "[s]tudents who plan to go to college to study the humanities or social sciences, those already intent on scientific careers, and those who may not wish to go to college at all..." [6]. And the historically-themed science courses one does see today are almost exclusively "distribution" courses intended primarily for non-science students. Teachers of science courses for future scientists, doctors, and engineers, though, have apparently never found a compelling argument for curriculum overhaul.

This is why I think it is important to stress—and to build some momentum behind—an alternative motivation: we should include history because an a-historical science course is necessarily to at least some degree a *dogmatic* science course, i.e., an un-scientific science course. If it is to present its subject matter accurately, science education simply must include not only science's conclusions but also the unique and rich process involving hypothesis, controversy, experiment, criticism, testing, debate, and accumulation of evidence that *makes* those conclusions scientific. Restructuring science education in this way would make science more accessible to more students, but that is merely a side benefit. The primary goal would be to teach science more *accurately*—not to dumb it down, but to keep it real.

Doing this should also yield enormous practical benefits. Students who were familiar with historical scientific controversies would gravitate toward (instead of being intimidated by) contemporary and future controversies. Students who have been inspired by the excitement of scientific puzzle-solving are more likely to engage with—and solve—the puzzles of tomorrow. Students who recognize that today's liberating technologies have grown out of scientific hypotheses that were initially derided as metaphysical and unscientific, will tend to be more courageous in fighting against present and future nay-sayers. And students who know both that many reasonable and empirically-successful ideas have nevertheless turned out to be wrong, and that there is such a thing as an unreasonable hypothesis that doesn't even warrant further consideration, are more likely to exhibit a proper scientific skepticism.

In short, we can and should speed and smooth our path to the future by refocusing science education around *historical scientific controversies and their resolutions*. This would, in effect, crowdsource innovation by putting a greater number of individuals in a much greater position to make the kinds of revolutionary discoveries that will uplift and liberate our descendants.

To flesh out this proposal to steer the future, let me give a couple of concrete examples of the kinds of historical episodes that I think should be highlighted in the science curriculum.

9.2 Ptolemy and Copernicus

The original scientific revolution was the proposal, by Copernicus, that the Earth was not the static center of the universe but was instead a planet—similar to Venus, Mars, and the others—which rotated daily and orbited the Sun yearly. According to the standard lore that most students absorb from textbooks and other sources, the geo-centric model of Ptolemy had been good enough for the ancients, but had required endless *ad hoc* fixes (such as the addition of ever-more epicycles) to conform to the data. By the 1500s, both the complexity and observational inadequacy of Ptolemy's model were out of control, and Copernicus thus adopted the only reasonable solution: he jettisoned the outmoded geo-centric model in favor of the elegant and observationally more accurate helio-centric theory.

But this historical account is utter nonsense. It is true that, by Copernicus' time, slight inaccuracies in the model parameters introduced by Ptolemy had caused a problematic cumulative drift, giving rise to a growing calendrical crisis. But any improvement associated with Copernicus' heliocentric theory was in fact a result of tweaking the fit parameters—not a result of making the Sun instead of the Earth the center of the system. Indeed, as far as the apparent positions of the planets are concerned, it is easy to see that there is a perfect one-to-one correspondence between the two theories.

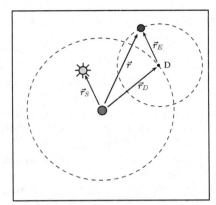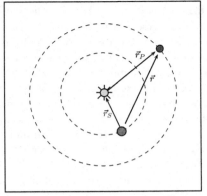

Fig. 9.1 The *left panel* shows the basic construction whereby the motion of a planet (say, Mars) is explained in Ptolemy's geo-centric theory. The "deferent point" D undergoes uniform circular motion around the earth, while the planet simultaneously moves around the epicycle centered at the (moving) point D. At any moment, the position $\vec{r}$ of the planet with respect to the Earth is the vector sum of $\vec{r}_D$ and $\vec{r}_E$. The *right panel* shows the corresponding construction in the heliocentric theory of Copernicus, where the relative position of the planet $\vec{r}$ can be understood as the vector sum of $\vec{r}_S$ (the position of the Sun with respect to the Earth) and $\vec{r}_P$ (the position of the planet with respect to the Sun). With the two planets (Earth and Mars) undergoing uniform circular motion, the perfect one-to-one correspondence between the two constructions—with $\vec{r}_E$ corresponding to $\vec{r}_S$ and $\vec{r}_D$ corresponding to $\vec{r}_P$, and hence the two $\vec{r}$s being identical—is clear

In particular, as indicated in Fig. 9.1, the motion of the Ptolemaic deferent point around the earth corresponds directly to Copernicus' motion of the planet around the Sun. Similarly, the motion of the planet on its epicycle in the Ptolemaic model corresponds directly to the motion of the Earth around the Sun in Copernicus' model. (For the inferior planets, Mercury and Venus, the correspondence is reversed.) So the two theories make identical predictions for the position $\vec{r}$ of the planet (with respect to the Earth) and hence the apparent position of the planet against the background of fixed stars (assumed very distant).

So far I have described only the "first order" versions of both theories. These provide a good qualitative description of the motion of the planets, but are not sufficiently accurate to account for the observational data as it was known already to Ptolemy. So both theorists, Ptolemy and Copernicus, introduced a variety of corrective devices including "eccentrics" (moving the appropriate body—Earth for Ptolemy and the Sun for Copernicus—slightly away from the circles' centers) and additional smaller epicycles. One of the corrective devices that Ptolemy had utilized—the "equant"— was regarded by Copernicus as an abhorrent departure from the basic principle of explaining heavenly movements in terms of uniform circular motions. Copernicus thus introduced additional epicycles to do the jobs that had been done by Ptolemy's equants.

Of course, Copernicus' system did have several virtues. Whereas the basic ("first-order") deferent-epicycle constructions for each planet were, in Ptolemy's system, independent, six of the circles (one for each planet, including the Sun) corresponded to a single circular motion in Copernicus' system (namely, the Earth's orbit around the Sun). So the helio-centric model fixed the relative sizes of the planets' orbits (and could consequently account for the observed variations in Mars' brightness) in a way that Ptolemy's theory didn't. And it also provided a natural explanation for something that was a sheer *coincidence* according to Ptolemy—namely that (as illustrated in the Figure with the identical vectors $\vec{r}_S$ and $\vec{r}_E$) the motion of each planet on its major epicycle (or, for Mercury and Venus, the motion of its deferent) was "locked" with the motion of the Sun.

Still, at the end of the day, Copernicus' theory was about as complicated as Ptolemy's, and gave no substantial improvement in predictive accuracy. Any preference between them would then have to be based on subtle judgments of "naturalness" or other "fuzzy" criteria including compatibility with the best available scientific theories from other fields. Proponents on either side of the debate should have recognized that neither side was in a position to claim their theory was conclusively established. The debate between geo-centrists and helio-centrists was, in short, a *legitimate controversy*.[1]

How was it eventually resolved? As it turned out, the first really convincing evidence for the helio-centric theory appeared several generations after Copernicus proposed it, and came in a form he could never have anticipated. Galileo, for example, used the newly-invented telescope to show that the Moon had mountains, the Sun had spots, and Jupiter had moons—thus undermining the idea of an absolute division between the central Earth and the surrounding perfect heavens. Galileo's telescopic observations also revealed that the planet Venus displayed a complete set of phases, like the Moon, and was hence sometimes *behind* the Sun and sometimes *in front* of the Sun. This was simply impossible in Ptolemy's theory, but clearly predicted by Copernicus'. Meanwhile, Kepler was demonstrating that the unprecedentedly accurate observations of Tycho Brahe simply could not be accounted for with *any* (reasonable) combinations of uniform circular motions, and that the planets must instead move (relative to the Sun!) in accordance with what we now know as Kepler's laws.

9.3 Dalton and Avogadro

Richard Feynman once described the atomic theory of matter as our most important discovery about the natural world:

> If, in some cataclysm, all of scientific knowledge were to be destroyed, and only one sentence passed on to the next generation..., what statement would contain the most information in the fewest words? I believe it is the *atomic hypothesis* (or the atomic *fact*, or whatever you wish

---

[1]Note that "was" and "legitimate" are crucial here. Controversies are eventually settled such that the issue is no longer (legitimately) controversial. And not all claimed controversies (or, for that matter, consensuses) are legitimate.

to call it) that *all things are made of atoms—little particles that move around in perpetual motion, attracting each other when they are a little distance apart, but repelling upon being squeezed into one another.* [1]

And of course every student of chemistry or physics today learns that matter is made of atoms. But most students do not learn anything about the way that the atomic theory arose or the kinds of evidence on which its fate turned.

The speculation that matter might be composed of imperceptibly tiny particles had a long history going all the way back to Ancient Greek philosophers. But the atomic theory that scientists accept today had its roots in the Law of Definite Proportions that was first championed by Joseph Proust in the 1790s. For example, suppose it is observed that 5 g of some element A can combine chemically with 8 g of some other element B to form 13 g of the chemical compound C:

$$5 \text{ g } A + 8 \text{ g } B \rightarrow 13 \text{ g } C.$$

Then the precise 5-to-8 ratio is characteristic of C in the sense that any sample of C must contain (and is in principle decomposable back into) A and B in this exact ratio. Thus, if one attempts to chemically combine, say, 5 g of A with *9 g* of B, the result will be the same 13 g of C as before plus one *leftover* gram of un-reacted B.

Around 1803, John Dalton pointed out that this empirical law could be explained if samples of each chemical element were composed of a large number of identical atoms. Then, for example, the reaction indicated above could be understood as some large number (N) of the following elementary atomic combinations:

$$\text{(A)} + \text{(B)} \rightarrow \text{(A)}\text{-(B)}$$

where the "(A)-(B)" on the right hand side stands for one molecule of the compound C. Note that if this is the correct description of the reaction in question, it follows that the individual *atoms* of A and B have masses in the ratio 5:8.

On the basis of his proposed atomic explanation of such reactions, Dalton in effect predicted the Law of *Multiple* Proportions: if the same two elements can combine to form two *distinct* chemical compounds, there should exist a small-whole-number ratio between the amounts of one element that combine with the same fixed quantity of the other. For example, suppose that our elements A and B can also combine chemically to form the compound D as follows:

$$5 \text{ g } A + 16 \text{ g } B \rightarrow 21 \text{ g } D.$$

The masses of B (8 g and 16 g, respectively) that combine with the same fixed 5 g of A in the two reactions are indeed in a small whole-number ratio (namely, 1:2) with each other. On Dalton's atomic hypothesis, this second reaction can be easily explained, as N instances of the elementary atomic reaction

$$\text{(A)} + 2\text{(B)} \rightarrow \text{(A)-(B)-(B)}$$

where the "(A)-(B)-(B)" on the right represents a single molecule of compound D.

Note, though, that in these sorts of cases an important ambiguity arises. We have explained the observable chemical reaction data, on the atomic model, by assuming that the compound C is *diatomic* (with D then being triatomic). This implies the 5:8 atomic weight ratio for A and B. But, for all we know, it could just as easily be the case that D is the diatomic molecule, in which case the relative atomic weight of A and B would instead be 5:16, and C would be triatomic! (On this scheme, a C molecule would look like this: (A)-(A)-(B).) Neither the relative atomic weights, nor the atomic composition of compounds, can be unambiguously determined from the reaction data alone.

And to add to the confusion, around this same time another distinct chemical combination law, pertaining to *gases*, was enunciated by Joseph Gay-Lussac. Gay-Lussac's "law of combining volumes" noted that, when two or more gaseous substances are involved in a chemical reaction, their *volumes* are in small whole-number ratios. For example, suppose that A, B, and C are gases and that the reaction above can be expressed (in terms of volumes, all measured at the same temperature and pressure) as:

$$10\,\text{L}\,A + 30\,\text{L}\,B \rightarrow 20\,\text{L}\,C.$$

Such a reaction can be easily accounted for in terms of combining atoms if we assume that the *number densities* of the different gases are in small whole number ratios. For example, suppose that the 10 L of gas A contains N A atoms. And suppose that each 10 L of gas B contains $N/3$ B atoms—i.e., suppose that the number density of A is three times the number density of B. Then the above reaction can again be understood as N copies of the elementary reaction

$$\text{(A)} + \text{(B)} \rightarrow \text{(A)-(B)}.$$

with the same implied 5:8 atomic weight ratio for A and B that we saw above. Finally, note that on this scheme there are N C molecules produced in the reaction, and hence $N/2$ C molecules per 10L of gas C. So the number densities of the three gases (A, B, and C) would be in the ratio 6:2:3.

But again there are ambiguities. As pointed out by Amedeo Avogadro in 1811, the above reaction involving gaseous A, B, and C could also be explained on the (beautifully simple) assumption that the number densities of all three gases are *equal*. One need merely abandon the (so far, tacit) assumption that elements are *monatomic*, i.e., that the smallest chemically meaningful particles ("molecules") consist, for elements, of single atoms. Avogadro would thus have explained the above reaction as N copies of the elementary atomic reaction

$$\text{(A)}-\text{(A)} + 3\,\text{(B)}-\text{(B)} \rightarrow 2\,\text{(A)}-\text{(B)}-\text{(B)}-\text{(B)}$$

where the "(A)-(B)-(B)-(B)" on the right represents a single molecule of C. Note that Avogadro's Hypothesis—that the number densities of all gases are equal—*requires* us to make both A and B diatomic: there is no way to divide one (A) or three (B) atoms evenly between two (C) molecules! And note that relative atomic weights can again be determined, but that they are very different. According to Avogadro, our 5 g of A contains $2N$ A atoms, while our 8 g of B contains $6N$ B atoms, so the atoms of A and B have masses in the ratio $(5/2){:}(8/6) = 15{:}8$, rather than the previously suggested 5:8 or 5:16 ratios.

Stepping back, the situation vis-a-vis atoms in the first decades of the 19th century was roughly as follows: the several empirical chemical combination laws could be naturally explained in terms of the chemical atomic theory, and these explanations seemed to bring important physical properties of the atoms (such as their masses) within reach of empirical determination. But there were *reasonable disagreements* about which way of determining these was right, and even about which way was the simplest. For example, which is simpler—the scheme in which A and B are monatomic and C is diatomic and their number densities are in the ratio 6:2:3, or the scheme in which their number densities are in the much simpler ratio 1:1:1 but A and B are both diatomic and each C molecule contains one (A) and *three* (B)s? The only honest answer is: it's not at all clear. Determining the relative atomic weights and molecular compositions (and hence assessing the chemical atomic hypothesis that gave these meaning) was controversial—and rightly so.

And as in our previous example, the controversies were only finally resolved, much later, and by totally unanticipated sorts of evidence. For example, it became clear by the 1850s or so that the kinetic theory of gases provided the correct explanation for the physical behavior summarized in the empirical laws of Boyle and Charles. But (coupled with the recognition that temperature is a measure of the average kinetic energy of the molecules) the kinetic theory dictated that different gases, under the same conditions of pressure and temperature, would have equal number densities. In addition, there was accumulating circumstantial evidence such as the 1819 law of Dulong and Petit, according to which a wide variety of substances had nearly identical heat capacities per atom—*if* one calculated these using the relative atomic weight assignments based on Avogadro's hypothesis. Similarly for Mendeleev's construction of the periodic table of the elements: the regularities involved in both of these cases would apparently represent fantastic coincidences if Avogadro's atomic weight assignments weren't right.

9.4 Past, Present, and Future

I think students should know not just that, according to current scientific authorities, the Earth goes around the Sun and matter is made of atoms. They should understand in addition something about how those ideas arose and why they were, when first proposed, controversial. They should know that Copernicus was dismissed as a "fool [who] wishes to reverse the entire science of astronomy" [7] and that Andreas Osiander, whom Copernicus entrusted to oversee the publication of his book, felt obliged to insert an unsigned preface urging the reader *not* to take Copernicus' ideas seriously, but to instead dismiss them as merely providing an alternative algorithm for making calculations: "it is the job of the astronomer to use painstaking and skilled observation in gathering together the history of the celestial movements, and then—since he cannot by any line of reasoning reach the true causes of these movements—to think up or construct whatever causes or hypotheses he pleases such that, by the assumption of these causes, those same movements can be calculated.... [I]t is not necessary that these hypotheses should be true[;] it is enough if they provide a calculus which fits the observations..." [8]

Students should similarly know that Feynman's "atomic *fact*" took nearly a century to be universally recognized as such. They should know, for example, that despite the apparent promise of the atomic theory, the ambiguities associated with (e.g.) atomic weight assignments led most scientists to dismiss the theory as useless and speculative for half a century. It then continued to be dismissed (with far less justification) as metaphysical and unscientific—by such people as Wilhelm Ostwald and Ernst Mach—into the beginning of the 20th century [9].

Why does any of this matter? Consider, for example, the contemporary state of quantum theory. By any honest assessment, the *physics* behind the quantum formalism remains completely unsettled. There are several distinct theories, which paint radically different pictures of the nature of the quantum world, but which remain empirically indistinguishable [10]. The situation, I think, is closely parallel to the historical examples we've been discussing: Ptolemy's and Copernicus' rival theories were, for a long period of time, also empirically indistinguishable, as were the competing schemes in the early 19th century for assigning relative atomic weights and molecular structures. A rudimentary knowledge of history would thus strongly suggest that we should pay careful attention to the issue of "interpreting" quantum mechanics, and expect unexpected innovations. In particular, we should expect that unanticipated evidence, coming from unanticipated directions, will at some point in the future resolve the ambiguity and allow us to finally discover, with certainty, the true physical meaning of quantum theory. In addition, we should expect that this resolution will open new, previously-undreamed-of doors in terms of technological applications—just as Copernicus' ideas initiated the path toward space exploration (and perhaps future colonization) and Dalton's ideas paved the way for computers and so much other contemporary technology with its roots in atomic and sub-atomic physics.

We should expect all of this, that is, *if we can resist the impulse to dismiss the controversy as "metaphysical" or otherwise meaningless and unscientific*. Unfortunately, though, the standard pedagogical orthodoxy on this particular controversy remains "shut up and calculate" [11]. Students, that is, are deliberately shielded from the existence of a controversy, and advised against wasting time thinking about it (should they somehow learn of its existence). I would not claim to know—and in general think it's foolish to try to guess—what technologies will transform the lives of our descendants. But it is plausible that the ultimate fate and practical dividends of something like, say, quantum computing, might hinge on understanding quantum *physics* correctly—just as so much modern technology depends on having eventually understood atomic weights correctly. But the innovations will never come—or will come only much later, after considerable pointless stagnation—if students are brow-beaten into dismissing the *most promising* questions as unscientific. It would only take a little knowledge of historical scientific controversies, for students to be able to see that the grounds for "shut up and calculate" in contemporary quantum theory are no different, in principle, from the attitudes expressed in earlier centuries by the likes of Osiander and Ostwald.

And of course, this isn't specifically about quantum computation or even quantum theory generally. Maybe the next big technological innovation will have nothing to do with micro-physics, but will instead come from astrophysics or chemistry or geology. Who knows! The point is that today's students—tomorrow's scientists and engineers—will be much more likely to gravitate toward the promising areas and then be much more likely to innovate once there, if they know something about how and when and why such innovations have occured in the past.

A science classroom that *highlighted* and *celebrated* historical scientific controversies would undoubtedly be more fun and interesting than a memorization-based, lecture-heavy classroom [12]. Science education is already moving toward more active-learning and inquiry-based approaches. Material about historical scientific controversies represents not only a vast untapped resource [13] for exciting labs and inquiry activities, but also a way of bringing the over-arching course content more in line with these existing methodological aims. It would make science more attractive to students. So more students would learn more science and bring scientific perspectives to their careers. But in addition—as I have tried to stress here—the controversy-focused science classroom would also produce students who are drawn toward, and equipped to resist dogmatic warnings to stay away from, the kinds of questions that have, historically, produced the most important revolutions in our thinking and technology.

In a world where science education focused on historical controversies, the road to the future would become a freshly-paved multi-lane super-highway, headed West. But to prepare ourselves to travel down that road, we need to do a better job of looking back and learning from the part of the road already traveled. That's why I say: back ... to the future!

References

1. Feynman, R.P., Leighton, R.B., Sands, M.: The Feynman Lectures on Physics, vol. 1. Addison-Wesley, Reading (1963) (Chapter 1)
2. Kuhn, T.: The Structure of Scientific Revolutions, pp. 136–137. The University of Chicago Press, Chicago (1962); Klein, M.: The use and abuse of historical teaching in physics. In: Brush, S.G., King, A.L. (eds.) History in the Teaching of Physics. University Press of New England, Hannover (1972); Whitaker, M.: History and quasi-history in physics education. Phys. Educ. **14**, 108–112 (1979); Allchin, D.: Pseudohistory and pseudoscience. Sci. Educ. **13**, 179–195 (2004)
3. McDermott, L., et al.: Physics by Inquiry. Wiley, New York (1995). Laws, P.: Workshop Physics. Wiley, New York (2004)
4. Norsen, T.: On the History and Future of Teaching Science Through History. AAPT Winter Meeting, Orlando (2014) (slides available from the author)
5. Millikan, R.A., Roller, D., Watson, E.: Mechanics, Molecular Physics, Heat, and Sound, Ginn and Company, Boston (1937); Taylor, L.: Physics: The Pioneer Science, Houghton Mifflin, Boston (1941); Longair, M.: Theoretical Concepts in Physics, Cambridge University Press, Cambridge (1984); Holton, G., Brush, S.: Physics, the Human Adventure, Rutgers, New Brunswick (2001)
6. James Rutherford, F., Holton, G., Watson, F.: The [Harvard] Project Physics Course. Hold, Rinehart, and Winston, New York (1970)
7. Luther, Martin, quoted in T. Kuhn.: The Copernican Revolution, p. 191. Harvard University Press, Cambridge (1957)
8. Copernicus, N.: On the Revolutions of Heavenly Spheres, translated by Wallis C., p. 3. Prometheus Books, Amherst (1995)
9. Brock, W.H. (ed.): Historical Overviews of Late-19th-Century Skepticism About Atoms Can be Found in: The Atomic Debates. Leicester University Press (1967); Nye, M.J.: Molecular Reality. Elsevier (1972); Chalmers, A.: The Scientist's Atom and the Philosopher's Stone. Springer (2009)
10. Bell, J.S.:Six possible worlds of quantum mechanics. In: Allen, S. (ed.) Proceedings of the Noble Symposium 65: Possible Worlds in Arts and Sciences, Stockholm, 11–15 August, (1986); Bell, J.S.: Speakable and Unspeakable in Quantum Mechanics, 2nd edn. Cambridge (2004)
11. David Mermin, N.: What's wrong with this Pillow? Phys. Today **42**(4), 9 (1989)
12. Welch, W.W.: Review of the research and evaluation program of harvard project physics. J. Res. Sci. Teach. **10**, 365–378 (1973); Abd-El-Khalick, F., Lederman, N.: The influence of history of science courses on students views of nature of science. J. Res. Sci. Teach. **37**, 1057–1095 (2000); Galili, I., Hazan, A.: The effect of a history-based course in optics on students' views about science. Sci. Educ. **10**, 7–32 (2001); Garcia, S., Hankins, A., Sadaghiani, H.: The impact of the history of physics on student attitudes and conceptual understanding of physics. In: AIP Conference Proceedings 1289, pp. 141–144. (2010)
13. Holton, G., Rutherford, F.J., Watson, F.: The Project Physics Course Handbook. Holt, Rinehart, and Winston (1970); Palmieri, P.: A phenomenology of Galileos experiments with pendulums. Br. J. Hist. Sci. **42**, 479–513 (2009); Chang, H.: How historical experiments can improve scientific knowledge and science education. Sci. Educ. **20**, 317–341 (2011); Besson, U.: Historical scientific models and theories as resources for learning and teaching: the case of friction. Sci. Educ. **22**, 1001–1042 (2013); Timberlake, T.: Modeling the history of astronomy: Ptolemy, Copernicus, and Tycho. Astron. Educ. Rev. **12** (2013)

Chapter 10
Recognizing the Value of Play

Jonathan J. Dickau

Play is the highest form of research.

Albert Einstein

Abstract For humanity to positively shape its own future, we must recognize the value of play as an essential activity for learning and creative expression. Cognitive Science researchers, Neuroscientists, and Educators, have told us this for a while, but lectures by top researchers in Physics stress that playful exploration is also crucial to progress in both experimental and theoretical Physics. Play allows us to learn and innovate. The value of play to research is greatly under-valued—compared to its benefits—by modern society. Given opportunities to playfully explore; anyone including students and scientific researchers will learn more, faster. Thus; encouraging play fuels innovation and progress—the engines of economic prosperity. Experts from all the fields above echo that observation, both in published works and in personal conversations or correspondence. To retain our sense of humanity and survive to shape the future, human beings must realize that play is every bit as essential as hard work is, to our growth as individuals and as a culture. For humans to positively shape our own future, we must exalt that which makes us human, and to do that we must recognize the value of play.

10.1 Introduction

What must human beings do, to shape the future in a positive way that helps us to assure our survival and avoid a dystopian fate? Can Science aid our cause, to help us create a futuristic utopian ideal instead? Can the progression of knowledge

J.J. Dickau (✉)
Poughkeepsie, NY, USA
e-mail: jonathan@jonathandickau.com

© Springer International Publishing Switzerland 2016
A. Aguirre et al. (eds.), *How Should Humanity Steer the Future?*,
The Frontiers Collection, DOI 10.1007/978-3-319-20717-9_10

and the growth of human knowledge about the universe and ourselves provide the means to uplift and unite the human race through understanding—as in the Star Trek vision of Sci-Fi pioneer Gene Roddenberry? The possibility for such a future remains open, but there is a danger we will undermine our capacity to engineer this outcome, unless certain trends are reversed. Science can help us create a positive future for humanity, but we must be willing to apply what we have learned more broadly, and to exalt the search for knowledge and the process of learning over the information learned and the specific insights gained. To do this; we must recognize the value of play. Researchers like Alison Gopnik [1] have observed in the playful activity of the youngest children, the emergence of sophisticated experimental protocols to isolate variables and reveal how things work—while they play with various arrangements of objects—which prompted her to call them "little scientists." We need to cultivate this scientific curiosity, and the playful mindset that supports it, not only for the young but for older folks too—especially in the innovative workplace and in academia. If we want adult researchers and developers to make great advances and discoveries, we must give them freedom to play. But before that; to properly educate our young people for careers in Science, Technology, Engineering, or Math, we must encourage them to playfully explore ideas and concepts—and not to merely memorize facts—because *this* is what helps them develop the mental acuity and problem solving ability which will allow them to succeed and excel.

Humanity can make great strides, and *could* create an idyllic future using what we now know. However; we *must* encourage a playful approach toward acquiring knowledge, and an appreciation of learning and knowledge for their own sake, to do so. While learning specific bits of information makes an individual fit for a range of tasks, it is more general knowledge that allows a person to move from one task to another as required by real-life circumstances. Humans routinely exceed the capabilities of machines in this area, but are expected more and more to function like automatons rather than humans. This expectation is now projected onto Education, and disturbingly placed on both teachers and students. To prosper as a race, however; we must exploit what makes us uniquely human and gives us the power to innovate. The innate intelligence of play is a wonderful path to understanding, giving humans the capacity to become scientists. At its core, Science *is* play! Scientific exploration is a ritualized extension of the playful exploration and experimentation of the very young. To a scientist; what we don't know about the universe inspires no fear, but offers a sense of awe and wonder like that of a child. Unfortunately; we have not learned how to nurture the behaviors that lead young people to become thinkers and innovators, because our culture encourages only those who are very good at Science to have the fun of working in that field. We need to communicate that is isn't all hard work or memorization, that Science is and should be fun, and that there is great value to a playful approach in subjects like Physics or Math—or in other areas of Research and Development.

At the frontiers; Science is less about facts—and more about how we learn what is real. In lecture after lecture; I have heard top experts in Physics—including Nobel laureates—expound on the need for an open-ended and playful dynamic, to assure research success or scientific advances. The expectation for a predictable outcome can kill progress in research, because the swiftest progress is often made when there is only an interest to see what nature is telling us, with no specific expectation of what we will find. Though everyone looks for predictable results and a good return on their investment, progress in research defies such expectations, and is stifled by them. Anton Zeilinger lectured at FFP11 in Paris that he once told his employers "If you want results, don't expect results," and spoke to the need to be playful about how we approach research. At the same conference, CERN theorist John Ellis told a story of a visit once from Margaret Thatcher where he was asked about his job, and he professed to working through pages of difficult calculations to predict what they would find, then hoping to see something else when the experiment was run. Of course; Mrs. Thatcher asked him "Wouldn't it be better to actually see what you predicted?" And Ellis replied "No, because that way we wouldn't learn anything interesting." And so it goes, because the rest of the world sees knowledge as a collection of facts and scientists see it differently. To them; knowledge is more like an endless progression of new discoveries and better understandings. The scientists among us see the scientific method and scientific knowledge as a way of learning about the world, and they retain a sense of openness and wonder that the rest of us have lost, but is greatly needed for progress.

When I say we need to recognize the value of play; I mean it is something we *must* do, to prepare today's young people for careers in STEM subjects, and the rest of society for working with the advanced technologies of tomorrow. But beyond this, we should understand that a playful atmosphere in research and development labs is a legitimate and responsible tool for progress. While prediction and control of outcomes is essential to other endeavors, trying to impose this mindset on scientific research will do more harm than good, because Science is not like Manufacturing or Construction—where the stages in a project's progress can be charted and timetables adjusted through the allotment of resources. In those endeavors; the unknown is the enemy, which creates uncontrollable uncertainties and prevents prediction of outcomes or adherence to timetables. Science treats the unknown in a fundamentally different way, because the things we don't know—that make reality scary for the rest of society—are exactly what makes the universe interesting, exciting, and fun, for scientists. The unknown is ultimately what drives scientists to pursue knowledge. They gain an advantage by studying what other people have learned, to benefit from other's past victories. However, scientists do not imagine knowledge to be a mere collection of facts that are concrete, lifeless, and unchanging, but instead they see knowledge as a living body of understanding—that helps us to shine a light on the nature of reality, and allows us to unlock its secrets. More importantly; those who enjoy the most success in Science are those who remain playful in the face of the unknown.

We find, especially in the sciences, that adaptive reasoning skills are more essential than information or knowledge in a fixed form. And yet; we seem hung up on teaching facts, or presenting information as facts, rather than realizing that the progression of knowledge demands a different approach. While many use the explosion of knowledge and the speed technical knowledge becomes obsolete as a rationale to teach students more and more facts in the short time available to educators; experts assert that students would be better served if more time was spent on teaching how to learn and less on memorizing facts. Unfortunately; what the experts know has not reached those who set the standards and oversee their implementation—leaving parents, students, and educators, scrambling to make up for what planners and administrators have not learned how to do effectively. It is true that students of today must learn more, and must learn faster, to graduate with the essential skills to function in modern society. But their success hinges on learning how to learn, and how to think for themselves. Nor can we imagine that the older generation is expendable, because they must convey a love of learning to the young—for young people to become inspired by the pursuit of knowledge for its own sake. One way we can create a better future is to encourage playful engagement with Science and Math, where we make it fun for all. If we can nurture the playful spirit all humans have as infants, and scientists need to advance human knowledge; this is how humanity can shape the future most positively.

10.2 Playful Learning Landmarks

Any learning process must proceed through stages, where basic knowledge acquired early on is then applied later in more complex settings. An important landmark in early childhood development is called object constancy. This is when we recognize that things are persistent, so they continue to have an existence even when they are out of sight or reach. For a very young child, a game of 'peek a boo' yields great pleasure, because it is a mystery every time the adult hides and a new discovery each time they emerge, but later on there is no mystery—since the constancy of objects and people is assumed. Once that bridge is crossed, though, a large number of other learning landmarks await as we discover how different collections of objects and people can be combined, or can interact. This is where play begins, and how our process of learning commences. Children play to figure out how things work, and to see how they are meant to go together or what is their function, and this is very much like the experimentation of scientists. Children are curious and they want to explore—to see different things, try different things, and go different places—learning how things change, and what stays the same. One important skill we must learn early on is how to navigate, therefore, and this leads to another learning landmark. Navigation at sea was made possible by a process called triangulation, and this same process is what allows toddlers to figure out how to get around—once they are mobile. As one moves, objects along the periphery grow as one gets closer and shrink as one moves away, so this allows us to make a determination of both our relative position and the sizes of objects.

One can make increasingly more accurate determinations, by learning the size of various landmarks and their distance from one another. But it all starts with a process of 'observe, explore, and compare'—repeated endlessly—where one comes to learn 'this is bigger than that' and 'this span is farther than that one.' Whether the landmark is a tower on shore for a ship at sea, or a refrigerator in the kitchen for a toddler taking his or her first steps, the process of triangulation is the same. Generalizing a bit, we see a process of dimensional estimation, or determination of the dimensionality of objects and our surroundings. This is something we must all learn, early on. Once this insight is acquired, however, something remarkable occurs. The research of Judy DeLoache [2] shows that children below age $2\frac{1}{2}$ display a 'dimensional confusion,' where they will attempt to put on shoes much too large or get into a toy car or chair that is much too small to sit in. They also have difficulty distinguishing 3-d objects from 2-d representations at that age. But once the developmental landmark is reached, that allows children to accurately estimate the dimensions of things around them, they also acquire an increasing ability to recognize and employ symbols, and to develop symbolic reasoning. My deduction, as noted in previous work [3], is that this ability to triangulate and to estimate sizes and distances is specifically what enables us to decode the symbolic realm, and to develop symbolic thinking.

What starts out with a literal cycle of observation, exploration, and comparison, becomes a process of systematic experimentation—where comparisons become more subtle as 'observe' and 'explore' take on much broader definitions. At first; a distance estimation may refer to the physical distance between two points in a room or a yard. But later; one can estimate the distance between abstract concepts in a multi-dimensional symbolic space, which represents the extended variables in the domain where those concepts have a specific range. Thus; Mathematics can be applied to ideas and relations between them—in the domain of pure thought—and one can use the same type of reasoning to understand the fundamental nature of physical reality through Physics. So when children playfully explore, and end up learning how to navigate by estimating and later grasping dimensionality (thus knowing 2-d from 3-d), they can learn about the language of symbols. This, of course, unlocks the door to all kinds of learning that was not possible before. In my view, this conceptualization is exactly like the brainstorm of Gerard 't Hooft, the Holographic Principle [4] relating 2-d and 3-d realities, which unlocked for physicists endless realms of undiscovered information, and opened new roads to further exploration and discovery. If this insight is put into perspective; it is perhaps like the advent of language, in terms of the transformational effect on our culture. It may take a while, considering how slowly the understanding from Relativity and Quantum Mechanics has filtered through into the general population. But maybe to future scientists; those two pillars of our present-day scientific knowledge will be relics from a period when Modern Physics was still in its infancy, before the 'Holographic Universe' Era.

10.3 Child-Like, Adolescent, and Adult Play

Play is the road to learning, but not all playing is the same. As Gopnik and her colleagues have learned [5], the play of very young children tends to be a mission of learning and discovery which is very much like the research of scientists. Pre-literate children spend a lot of time learning about the nature of reality—discovering how things work by testing their theories. But later development brings more interactive forms of play, because the capacity for interaction increases as we move further along nature's neurological and physiological timetable. As physical development brings more complex and sophisticated neurological structures online, the ability becomes available for more complex and sophisticated forms of social play. The recent work of Joseph Chilton Pearce and Michael Mendizza [6] emphasizes the value of play for learning at all ages, and traces how the forms taken by playful activity evolve over time, and with the development of our cognitive faculties. How we learn, at each stage of development, is hard-wired by specific patterning of natural development— but while this generally happens at a specific age or a distinct range of ages, it is uniquely dictated by the neurological development of each individual, on a case by case basis. This suggests that we need to tailor our instruction to what children are primed to learn at each stage of their neurological growth, rather than imposing an external timetable on them.

My own talk at FFP11 in Paris [7] detailed how understanding the structure of the brain and the changing nature of play for different age groups aids both success with Science and Math instruction and research progress. While the play of young children tends to be amicable, at least with some adult supervision, the play of adolescents takes on a different character—especially for males—becoming more competitive as the teen years progress. And while modern society celebrates competitiveness; this adolescent form of play is *not* the final stage in our development, and competition is *not* what fuels our greatest and highest accomplishments. The growth spurt in the brain initiated at puberty brings the deepest portions of the brain to their final stage of development, while beforehand and afterward the emphasis is on develop-ing the structures that support higher cognition and abstract reasoning. Therefore; while child's play and mature or adult play emphasize higher-brain function, which is expressed in development of the neocortex, adolescent play is more involved with the activity of lower-brain centers. After its final growth spurt; the action of the mid-brain or hindbrain is automatic, however, and there is no capacity for any additional learning, reprogramming, or higher cognition. The character of its reasoning is prim-itive or primal, and it has been called the 'lizard brain' because that is how it thinks. While over time our 'lizard brain' can be retrained, it appears to be fixed shortly after puberty because it responds to change so very slowly. This is why it is important to move beyond adolescent play, and to emphasize activities that involve or activate the higher centers of the brain.

While it is not obvious; play is the most cerebral activity of all, in all its forms except adolescent competition, because an attitude of playful exploration stimulates activity and learning in the neocortex—the very highest region of the brain—which supports the most sophisticated types of reasoning. If we wish to reap the fruits of cerebral activities, we need to curtail activities that force us to use the 'lizard brain,' and emphasize those that allow us to use the neocortex instead. When people are intimidated into compliance, or compelled to adhere to an artificial timetable, their ability to make progress suffers. While necessity *can* foster innovation, often the best scientists can do is create the ideal conditions for a discovery to be made, and then wait for nature to reveal herself in the experimental results. Even in the face of extreme need, it is better to remain playful—tossing ideas around in one's head—than to become focused on how important it is to get the job done. That is; the 'lizard brain' cannot help us to innovate any faster, and struggling to work harder will not reveal the answers quicker either, because it activates a portion of the brain that is not up to the task. Instead of working harder; we need our researchers to be more playful, and should reward them for doing things in a way that allows use of the highest centers in the brain and facilitates higher reasoning. So; we must coax people away from competitive adolescent play to more cooperative adult play, and a 'win-win' mentality, to keep things cerebral.

10.4   Teaching Lifelong Learning

The notion that knowledge has value for its own sake is unpopular these days, as applying one's education toward finding employment in your chosen field is the paramount concern. Everyone is looking for a good return on their investment, and the entire field of Education—from pre-school through graduate school—is compelled to create measurable value in terms of employability. But I question whether this industrial vision of education serves the needs of our young people, or delivers the knowledge they need and the skills they require to use it. Teaching only information, delivered in pre-digested allotments, robs students of opportunities to learn that growing brains and nervous systems require. Compelling teachers to teach what is on the test first, and only later to convey ideas and concepts, frustrates the natural process by which learning occurs—because things 'want' to happen in the reverse order. In a local lecture; Education author Alfie Kohn told how one class learned to measure on its own, through a process of guided discovery by which they developed their own units of measure—introducing units and measures in a way those students will never forget. What kids learn through playful exploration is retained indefinitely. But we must assure that this fact known by Education researchers and innovative educators is shared broadly enough to be helpful. Speaking with Kohn after his lecture; I informed him that the same rules of learning he emphasizes for students and educators also work for researchers at the frontiers of Science, and he had not heard this before, though it came as no surprise. But we need make sure such knowledge is more broadly available, or more widely known.

While I appreciate the need to educate our young people, and the fact that they are our hope for the future; I feel that lifelong learning is too often neglected or left to chance, while we focus on the young. If we view educating our adult population as less essential than educating kids, we are robbing both our children and our adults of the learning experiences they need to usher in a better future. Without a well-educated adult population, our children will not have the learning opportunities to create the kind of world we desire—regardless of the quality of the education they receive while at school. When the adults at home are sharper and more knowledgeable, this encourages young people to learn more, while uninformed adults tend to make learning difficult for kids. Unless parents appreciate the need for education, and can assist in their children's learning process when not at school, the prospects for a bright future diminish—because essential skills are never imparted. We are quick to assume the skill set of many adults has become outdated, in modern times, because it is believed that all knowledge has a half-life—a limited range of applicability—before whatever is learned becomes obsolete and therefore inconsequential. This is only a half-truth however, as some knowledge is enduring or universal, and there is evidence the survival of an older generation and the wisdom of elders in our culture allowed humanity to escape doom on several occasions already [8], from the pre-historic past until today.

My recently departed friend Pete Seeger was a playful-minded fellow who remained sharp and continued actively learning, well into his 90s. If you asserted that he had no useful knowledge to impart at an advanced age, because everything he learned in school was obsolete, quite a few people would tell you otherwise. He and his wife Toshi, who passed last year, were passionate advocates of lifelong learning, supporting Science and Math education for girls as well as boys, for young and old people alike. But I've also been privileged to interact with Professors Emeritus, and other elders of academia who still have active minds—with a lot to teach and a passion to make their point—at an advanced age. Frank Lambert, now 95, led an effort after retiring from teaching, to reform the treatment of Thermodynamic Entropy in Chemistry [9]—moving it away from the notion that entropy is disorder, and toward a metaphor of energy dispersal or spreading—where now around 90 % of Chemistry textbooks have dropped the disorder metaphor. Steven Kenneth Kauffmann, who is 'only' 75, continually amazes me with a steady stream of new ideas that upgrade my understanding of Physics and challenge my intellect, in papers [10] and correspondence full of keen insights. So when I see colleagues in their 50s (my own age) or 60s marginalized, because too much of their knowledge is outdated, I have to wonder if those who determine this have any understanding at all. When I had questions about Decoherence and H. Dieter Zeh took time to correspond with me it was priceless, for example, because I got my answers directly from the world's foremost expert, and his advanced age was no issue.

10.5 Concluding Remarks

Talking about play; most people think of it as a way to waste time or enjoy some time off—when they are not working. Few get paid to play, in our society, and many of those who do are either involved in a competitive sport, or are musical performers, actors, and other entertainers. But if Science is to help us to create an idyllic future as in the Star Trek universe; we must appreciate the role that play takes for scientists, mathematicians, inventors, and other innovators. Sure, there is serious work involved, and one must get every detail exactly right—before one can establish the working conditions where the playful phase of the exploration process provides dividends. However; playful researchers make more discoveries, and win a larger portion of the accolades in Science, than their more timid and conventional peers. Speaking at FFP10 in Perth; Nobel laureate Doug Osheroff explained that researchers must be willing to question the wisdom of today's theorists, and to look in unexplored regions of the parameter space, in order to discover new things. His talk on "How Advances in Science are Made," which he has delivered in several venues [11], was full of examples of how scientists must retain a sense of play in order to make progress. It seems clear that, in the realm of scientific research, the spoils go to the playful rather than the methodical.

In today's world, where a guaranteed return is a requirement for investment of resources and everyone is scrambling for a piece of a shrinking pie, the need to provide an open-ended environment to researchers—to foster progress—is often forgotten. Furthermore; now when researchers need more opportunities to play and explore—to accelerate progress—we force them to deal with tighter and tighter restrictions, and this slows the pace of progress instead. Efforts to have scientists conform to the norms of prediction and control—favored by administrators—are doomed to backfire, because these misguided efforts fail to grasp the fundamental nature of research, and the standard methods are designed for tasks that are very different. Yet increasingly; researchers face a situation where, upon entering the workforce, they are burdened with heavy workloads and administrative duties—that take them away from their research, and slow their progress [12]. When we send our most able scholars the message that it is not OK to play and they must do 'serious work' instead, we are doing them and our world a disservice. They are the ones who will create the idyllic future that Roddenberry envisioned, if we are to see one at all. We should be celebrating scholarly achievements to as great a degree as we do those of athletes on the field! Perhaps more importantly; we should revere new knowledge once it is received, because seeing great scholarly accomplishments like Perelman's proof of the Poincaré conjecture [13] shows us the inherent worth of such pursuits. Of course; a full appreciation of the importance of that work would require a much more well-educated general population.

The challenge, then, is to inspire more people to seek higher education, to make Math and Science more fun to learn, and thus to elevate the general intelligence of the populace, in the core STEM subjects. To do this; we must acknowledge that these are playful pursuits by nature, and make it OK for scholars in these fields to actively play. Play is far more universal, being the root source of *all* learning, and indeed of all consciousness and cognitive intelligence, but it clearly finds expression in these subjects. While Math and Science are full of hard topics to learn; they are, at their heart, fun! But this is only one reason I say that Science is play. The very best scientists, those at the forefront of their field, seem to have a defining characteristic in common; they retain a sense of play, delighting in the awe and wonder of the natural order. And this is something all young humans display at an early age. For us to be the creators of a better world; we need to nurture and cultivate the playful spark of curiosity we are all born with, which is a defining characteristic of all human beings. Knowledge is worth having for its own sake, apart from any financial benefit it might confer. However; the kind of knowledge scientists seek is not a collection of facts, but a living, breathing thing. Science brings us a kind of knowing that is dynamic and endlessly expands the boundaries of knowledge. It is not a commodity that can be contained and retained, but rather it is a playful never-ending voyage into ever-increasing understanding and intelligence. This is what we need most, to create the idyllic future we want—just as Roddenberry envisioned. Play is the most fundamental freedom for children, and it must be preserved, but giving Science-minded adults freedom to play will help humanity reach the stars.

10.6 Reflections and Observations

Having the opportunity to comment further here, after learning a panel of experts chose this essay for a prize, brings a kind of validation that my message was heard. However; I know impressing scientists with the idea that a playful attitude has value for learning and research is like 'preaching to the choir.' The tough thing is selling the idea to business people, economists, and financial gurus, that letting scientists and developers play more freely is the surest road to increased innovation and progress. But it is the honest truth! Playful exploring allows us to discover and develop things that no amount of memorized information can yield. You can see a gleam in the eye of the most successful researchers that shows they have not forgotten how to play or why remaining playful about coaxing nature to reveal her secrets can yield swifter results. Young people need to learn that doing Science can be a lot of fun, and that folks who explore the frontiers of knowledge lead interesting and exciting lives. Having heard him speak; I imagine Quantum Physics researcher Anton Zeilinger goes to work in the morning thinking "this might be the day I learn something nobody else has seen before." That kind of job incentive is a powerful motivator for progress, and it is something palpable for all the top researchers—a fact that seemingly eludes those who see progress solely as the incremental product of their hard work. Progress needs thinkers and innovators, and those people need the freedom to look beyond predictable outcomes or safe assumptions—to playfully explore where no one has gone before—in order to create a brighter future.

People working in Science, and innovators working to create the technologies of tomorrow, *must* see things differently—or need to understand there is more to the story. While some things decidedly *are* the incremental product of work done over time, research and development are fundamentally different. The hard work comes mainly during the phase of preparation, where all the tools and resources necessary are assembled and utilized—to create the pre-conditions for the real action to begin. Then comes the stage of the process where hard work is no longer enough, and one needs to have the ability to play well—to go beyond that point. It matters little what the area of specialization might be, within subjects like Physics, Chemistry, and Biology; the same rules apply. Experimentalists play with equipment and parameters, or samples of various kinds, while theoreticians play with ideas, but the playful nature of their activity is very much the same. Different possibilities must be examined, from every possible angle, to ascertain what is, might be, or definitely is not true. And only when a number of different models or interpretations are sorted into such categories is there an increase in our knowledge and understanding over time. To 'toss ideas around' in one's head is an essential skill for any scientific researcher or technological developer, but this is fundamentally different from the kind of thinking that is directed toward a specific goal, or uses a single standard method to solve a well-understood problem—because many different things must be tried, and you don't know exactly what will work best.

Play is essential to many kinds of endeavor, and to achieving the highest levels of performance within that field. It is not the absence of work, as some would have us believe, but instead; play is often very hard work, yet it is joyfully undertaken, in the spirit of exploration. This is what fuels humanity's dreams of progress, and leads us into the stars. The wide-eyed sense of wonder and awe, which is common for children, is unfortunately uncommon for adults. However; many adults I've met who retain this quality are scientists and developers, people who explore the frontiers and use their brains in their daily work. And I can state unequivocally, that those who most exude this childlike sense of wonder are the people at the very top of their respective fields, who have made significant discoveries or important contributions to Science and Technology. So it is obvious that being playful works, to foster innovation, speed the process of discovery, and boost progress. Our institutions and society need to respect and endorse the efficacy of play, in all creative endeavors, rather than perpetuating the myth that play is not work, or pretending it is not essential to the highest levels of performance. As a culture; we need to become much more like little children—by allowing and encouraging adults to be playful in their pursuit of knowledge—if we are to gain the skills that will enable us to ascend to the stars.

If we want to see a future that resembles the one in Gene Roddenberry's 'Star Trek,' we need to foster a culture of lifelong learning, by sending a message to both young and old that the pursuit of knowledge is a worthwhile goal. Solving the problems that face modern humans requires a commitment to using our intelligence to conquer them, but if we rise to that challenge; the entire universe is ours to explore! Only the unending increase of knowledge will bring us to the fruition of that dream, but it might also lead us to a future even brighter than Roddenberry could have imagined. The thing is; we need to begin moving in that direction soon, if humanity is to get there at all. By encouraging the playful exploration of possibilities and the playful pursuit of knowledge—which values what we don't know as much as what we know—we allow the discoveries to come sooner, and foster a greater understanding overall. The truth is; there is far too much shallow reasoning in today's world, and this is part of why we face some of our most vexing problems. Ergo; if we want to turn that around, we need to foster and inspire deep understanding—which contains or embodies the power to solve problems. Only a truly deep understanding can solve the most difficult problems we face, and a playful approach to learning can make that understanding possible. The idea that play is expendable comes from the mistaken notion that knowledge is a collection of facts which can be memorized, but memorization does not confer the same problem solving ability as playful learning. This is why I assert that we must recognize the value of play, in order to create the bright future we desire.

Endnotes

Dimensional Estimation Through Triangulation

Triangulation—the ability to triangulate, to navigate or to determine the size and distance of objects, depends on perspective—as generalized in Projective Geometry—but the basics are encapsulated in Trigonometry, the study of triangles. Using 'observe, explore, compare' one could note that a lighthouse tower on shore at sea is as big as one's fingernail at arm's length—when it is first sighted—and as large as the entire finger at arm's length—once one moves closer (as in Figs. 10.1 and 10.2). Using the properties of right triangles; we can calculate how much closer we are, or even exactly how far away—if the angles of elevation have been measured precisely and it is a landmark of known size. But this essential skill for navigators is acquired at an early age by every child, in the process of their learning how to gauge the dimensionality of objects and the environment.

The most basic relation in Trigonometry is called the Pythagorean Theorem, which states that $c^2 = a^2 + b^2$, where c is the hypotenuse, and a and b are the legs of a right triangle (Fig. 10.3). This formula allows us to calculate the length of any side, knowing the other two, and given that the angle between a and b is a right angle. It is almost as simple to find the unknown distance, given one side and an angle. If we know the height of the tower (which stands at a right angle) and measure the angle of elevation Θ, we can calculate our distance from the tower using the formula $\tan \Theta = \frac{height}{distance}$. This allows our estimates to be made precise.

Fig. 10.1 A lighthouse tower appears smaller at a distance

Fig. 10.2 The same tower appears larger when closer

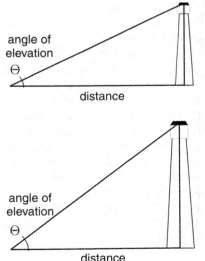

Fig. 10.3 The Pythagorean
theorem gives any side of a
right triangle, if we know the
other two

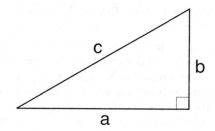

Ranging—the process of dimensional estimation requires calibration, in order to be effective. We must learn how big things are. Very young children display 'dimensional confusion' when experimenting with the calibration of their grids, to determine what is 'close enough' to work and what fails to match their needs or expectations. Children above the age of $2\frac{1}{2}$ lose this 'confusion' and display increasingly more ability to distinguish the dimensions of objects and their background environment correctly. In addition to estimating size and distance, children learn to tell the difference between 2-d surfaces or images and 3-dimensional objects, as well. This is one of the key factors that sets the stage for the acquisition of knowledge using symbols, and for symbolic reasoning, in human children.

Playful Comments

Michael Mendizza commented (after reading an earlier draft):

> You are circling around the tip of a galactic iceberg.
>
> Consciousness is play. Thought is play. To treat thought and consciousness any other way is to 'play falsely,' pretend that thought-consciousness is not what it is, which is a form of self-deception and shared delusion.

And he continued with these words:

> Personally I question pinning so much of your thesis on science. Humanity, sanity, appropriate and sane social orders, kindness, the ability to see 'what is,' which is the essence of science and also what contemplative traditions call enlightenment, is much more fundamental. All of this critically depends on appreciating that play, Maya, is what thought and consciousness is. To not see this is to live in delusion, which we do. Play liberates us from 'playing falsely' with thought and consciousness.
>
> Play is also the gymnasium of imagination, the place where we develop our capacity to create, which mirrors and is creation itself. The enlightened use of imagination is causal, literally we are the image and likeness of creation (God if you must), but playing falsely with thought consciousness means that what we create is distorted, and therefore we become the enemy. We are the enemy because we don't understand the true nature of what consciousness-thought really is. Play!—Michael Mendizza (on 1/31/14)

Playful Learning Resources

There is such a wealth of information about play available, that my repeated attempts to collate the relevant sources have only increased the number I found. I should start by recommending the books and articles of several authors I cited, especially Alison Gopnik, Joseph Chilton Pearce, and Michael Mendizza. Of course; books by Richard Feynman like "Surely you're joking.." and "What do you care what other people think?" contain plenty of insights on how a playful attitude benefits learning in Physics, but Michael Mendizza heartily recommends the works of David Bohm, as well, for deeper insights into how play is integral to learning and thinking. He also introduced me to the work of Dr. Stuart Brown, whose book "Play: How it Shapes the Brain, Opens the Imagination, and Invigorates the Soul" reinforces all of the messages in this essay, and provides additional insights on how play is essential to a broad variety of activities. The following links may also be helpful.

http://www.nifplay.org The National Institute for Play—founded by Stuart Brown M.D.

http://www.ted.com/talks/stuart_brown_says_play_is_more_than_fun_it_s_vital A TED talk by Dr. Brown "Play is more than just fun"

http://ttfuture.org Touch the Future—a project of Michael Mendizza with a team of experts

http://www.nurturing.us The Nurturing Project—another effort of Michael Mendizza

http://www.journalofplay.org The American Journal of Play—a multi-disciplinary journal devoted to the study of play. It has an impressive collection of papers stressing the importance of play to learning, as well as documenting its role in establishing a healthy society.

And finally; I am assembling my own collection of work on this subject, which will feature additional links to content found on the web, emphasizing the importance of play to Science.

http://www.scienceisplay.org Science is Play—a project of Jonathan J. Dickau

In closing; as my departed friend Ray Munroe would say,

Have Fun!

References

1. Gopnik, A.: How babies think. Sci. Am. **303**, 76–81 (2010)
2. DeLoache, J.: Mindful of symbols. Sci. Am. 60–65 (2005). Becoming symbol-minded. Trends Cogn. Sci. **8**(2), 66–70 (2004)
3. Dickau, J.: How can complexity arise from minimal spaces and systems? Quantum Biosyst. **1**(1), 31–43 (2007). Cherished assumptions and the progress of physics, 2012 FQXi essay contest entry, also published in Prespacetime **3**(13)
4. 't Hooft, G.: Dimensional reduction in quantum gravity, essay dedicated to Abdus Salam. October 1993. arXiv:gr-qc/9310026

5. Gopnik, A., Sobel, D., Schulz, L., Glymour, C.: Causal learning mechanisms in very young children.... Dev. Psychol. **37**(5), 620–629 (2001). Gopnik, A., Schulz, L.: Mechanisms of theory formation in young children. Trends Cogn. Sci. **8**(8), 371–377 (2004)
6. Pearce, J.C., Mendizza, M.: Magical Parent, Magical Child. North Atlantic Books, Berkeley (2003); Pearce, J.C.: The Biology of Transcendence. Park Street Press, Rochester (2002)
7. Dickau, J.J.: Learning to Cooperate for Progress in Physics, FFP11 talk slides at: http://www.jonathandickau.com/FFP11docs/LearningtoCooperateforProgressinPhysics.pdf proceedings paper at: http://www.jonathandickau.com/FFP11docs/JDickauFFP11.pdf or indexed at AIP: http://scitation.aip.org/content/aip/proceeding/aipcp/10.1063/1.4732721
8. Caspari, R.: The evolution of grandparents. Sci. Am. **305**(2), 44–49 (2011)
9. Lambert, F.: See http://www.entropysite.oxy.edu, http://www.secondlaw.oxy.edu, and http://www.2ndlaw.oxy.edu for details, links, and many examples
10. Kauffmann, S.K.: See http://www.vixra.org/author/steven_kenneth_kauffmann, and http://www.arxiv.org/find/all/1/au:+Kauffmann_Steven_Kenneth/0/1/0/all/0/1, for recent papers. Also see the FQXi forum discussion here: http://www.fqxi.org/community/forum/topic/1586
11. Osheroff, D.: How advances in science are made; find the slides for this talk at: http://www.stanford.edu/dept/physics/people/faculty/osheroff_docs/06.04.21-Advances.pdf, and video at: http://www.gallery.ntu.edu.sg/videos/v/nobel/osheroff/
12. Gibney, E.: 'Extreme' workloads plague scientists at the start of their careers, Nature News. doi:10.1038/nature.2014.14567 22 February 2014
13. Perelman, G.: Ricci flow with surgery on three-manifolds, arXiv:math/0303109; Finite extinction time for the solutions to the Ricci flow on certain three-manifolds, arXiv:math/0307245

Chapter 11
Improving Science for a Better Future

Mohammed M. Khalil

Abstract Science is the reason humanity reached this stage of progress, and science is humanity's guide to the future. However, to enable science to guide us to a better future, we need to improve the way we do science to accelerate the rate of scientific discovery and its applications. This is important to find urgent solutions to humanity's problems, improve humanity's conditions, and enhance our understanding of nature. In this essay, we seek to identify those aspects of science that need improvement, and discuss how to improve them.

11.1 Introduction

During the first half of the twentieth century, two scientific revolutions took place: *relativity* and *quantum mechanics*. They had a huge impact on our understanding of the universe, and led to many technological advances.

Relativity revolutionized our understanding of space, time, mass, and gravity. This understanding made many technological applications possible, such as particle accelerators, nuclear power plants, and the GPS.

Quantum mechanics revolutionized our understanding of particles and waves. It tells us we can only know the probability of finding a particle in a certain state, thus destroying the notion of a deterministic universe. Applications of quantum mechanics are numerous, such as the transistor, the laser, and the Scanning Tunneling Microscope (STM).

The advances in physics and technology changed our views about the universe. What was thought to be nebulae in our galaxy turned out to be other galaxies with billions of stars. The notion of a static universe became an expanding one that began with a big bang 13.8 billion years ago. Everything we observe in the universe turned out to constitute only 4.9 % of its contents; the rest is dark matter and dark energy.

M.M. Khalil (✉)
28th Omar El-Mokhtar Street, Alexandria 21914, Janaklees, Egypt
e-mail: moh.m.khalil@gmail.com

© Springer International Publishing Switzerland 2016
A. Aguirre et al. (eds.), *How Should Humanity Steer the Future?*,
The Frontiers Collection, DOI 10.1007/978-3-319-20717-9_11

Similar revolutions appeared in chemistry and biology: polymers changed our everyday products, medicine eradicated many diseases, the green revolution in agriculture saved us from starvation, and the discovery of the DNA revolutionized our understanding of life.

Science created wonders that would have been unimaginable a hundred years ago; still, wouldn't it be interesting to speculate on the wonders science will create in the future?

11.2 Can We Predict the Future?

The human race has always wanted to control the future, or at least to predict what will happen. That is why astrology is so popular.

Stephen Hawking [1, p. 103]

Many books were written about the future, and many people have speculated about the future, some of their predictions came true, but many did not.

One of the famous predictions that came true is Richard Feynman's 1959 lecture entitled: "There's plenty of room at the bottom", in which he considered the possibility of manipulating individual atoms [2]. Feynman's prediction came true with the invention of the STM in 1981, and his lecture marked the beginning of nanotechnology.

By contrast, a prediction that failed to come true is that of John von Neumann in 1948 about computers [3, p. 116]: "It is possible that in later years the machine sizes will increase again, but it is not likely that 10,000 (or perhaps a few times 10,000) switching organs will be exceeded..." The *transistor* made it possible to put more "switching organs" in less space; now, one can buy a computer with a billion transistor.

Science fiction novels, especially those of Jules Verne and H.G. Wells, also contain many technological inventions that came true in the future. In fact, almost all future predictions are about technology. However, considering the huge impact of science on our lives, we cannot successfully predict the future without taking into consideration the impact of science. Is predicting science possible?

To understand this question, let us ask another one: did anyone predict relativity, quantum mechanics, or the DNA? No, scientists can predict the *consequences* of science, but not the *scientific knowledge* itself. Physicists predicted the LHC would discover the Higgs boson, but they did not predict its theoretical "discovery" in 1964.

Predicting the content of new scientific knowledge is logically impossible because it makes no sense to claim to *know* already the facts you will *learn* in the future. Predicting the details of future technology, on the other hand, is merely difficult.

Eric K. Drexler [4, p. 39]

11.3 Choosing the Path to the Future

Science is the reason humanity reached this stage of progress; hence, we can expect that science will shape our future as well. However, the impossibility of predicting science renders any attempt to imagine the future incomplete at best. Thus, we cannot steer the future towards a certain vision. Instead, we can choose the *path* that leads to the brightest future. I believe *science* can lead us there, but we should make the conditions ideal for science to do so.

We should improve the way we do science to accelerate the rate of scientific discovery and its applications. Humanity is faced with many problems that need urgent innovative solutions; science is our guide to find those solutions, and improve humanity's conditions.

In the following subsections, I will try to identify those aspects of science and technology that need improvement, and discuss how to improve them to ensure accelerated scientific and technological breakthroughs.

11.3.1 Transcending Traditional Disciplines

During the renaissance age, science was not divided into disciplines. Disciplines emerged gradually when knowledge increased to enable researchers to study one subject in greater depth, and reach new findings more quickly.

However, nature is a whole that recognizes no disciplinary boundaries; so when knowledge increased, the need arose to combine knowledge from two or more academic disciplines to explain certain phenomena. This is known as *interdisciplinary* studies [3, p. 123], and it is getting increasingly important. This is evident from the number of interdisciplines that appeared recently, such as biophysics, astroparticle physics, nanoscience, and systems science. There are also *transdisciplinary* fields of study that use many disciplines in a holistic approach, such as environmental science.

It is hard for one person to know enough about two disciplines to do research in both, so inter- and transdisciplinary studies usually take the form of *collabora-tions* between scientists from various disciplines. Collaborations, however, face some challenges: scientists use different methods in each discipline, which can delay the progress of their project. Also, insufficient knowledge of each other's discipline can lead to misunderstanding between them. Further, scientists are usually unaware of problems faced by other disciplines, which can hinder starting collaborations in the first place.

The increasing importance of collaborations between scientists and engineers requires that undergraduate and graduate students be taught how to communicate and collaborate with researchers from other disciplines. Also, before starting new collaborations, researchers should acquire general knowledge about the other disciplines. In addition, university departments should give periodic talks on the problems they are working on to stimulate discussions with researchers from other disciplines, thus opening the possibility of interdisciplinary collaborations.

A great example about the importance of disregarding disciplinary borders is the *MIT media lab*. The lab is based on the idea of creating an environment for researchers from various disciplines to work together to change the world. The lab contains 25 research groups that disregard traditional disciplines. For example, the consortium Things That Think includes computer scientists, product designers, biomedical engineers, and architects working on digitally augmented objects and environments. The lab gave rise to more than 80 companies, and to many commercial products ranging from electronic ink to CityCar [5].

11.3.2 Creating New Specializations

Most engineers who work on renewable energy are mechanical or electrical engineers who decided to apply their knowledge to renewable energy. Wouldn't it be more effective if there were an engineering specialization on renewable energy that starts from undergraduate study?

Most of humanity's problems require knowledge from many disciplines. Collaboration is one way to solve them; another is creating *new specializations* designed specifically towards those problems. There are also emerging fields of study that has the potential of changing our future, such as nanotechnology, biotechnology, photonics, and artificial intelligence. Letting students specialize in those fields early in their study can accelerate the rate of innovation.

An example of this is the undergraduate majors of the recently inaugurated University of Science and Technology in Egypt. These majors include nanoscience, renewable energy engineering, and environmental engineering. In addition, every major has specializations; for instance, environmental engineering includes the specializations: climate change, water recycling, waste recycling, and water desalination [6]. I believe graduates from those, and similar, programs will be more equipped to excel in their field of study.

11.3.3 Big Science Versus Small Science

In the past few decades some *'big'* science projects appeared, such as giant particle accelerators, space exploration programs, gravity waves detectors, and the genome project. Those projects grasped the public attention, and rightly so, since they added a lot to our knowledge. Will the future of science be based on increasingly bigger projects? Or will *'small'* science projects contribute more?

The problem with big science is the cost. This is clear from the cancellation of the Superconducting Super Collider in the US, which would have been the largest particle accelerator on earth. Its cost increased from $4.4 billion in 1987 to $11 billion by 1993, and since foreign sources of funding could not be found, the project was cancelled [7].

Because the big projects are likely to become fewer and slower while the small projects stay roughly constant, it is reasonable to expect that the relative importance of small projects will increase with time.

Freeman Dyson [3, p. 125]

Does that mean we should do small projects only? No, the problem with big projects can be solved by *international collaborations*. The best evidence is the International Space Station that cost over €100 billion, according to ESA [8], making it arguably the most expensive object ever constructed. This cost was split among the 14 participating countries.

Unfortunately, International projects can be hindered by political conflict between nations. To overcome this difficulty, governments and funding agencies should allocate a certain sum of money to international projects, and form a committee, mostly of scientists, to decide on which projects they should collaborate, disregarding any political conflict.

11.3.4 The Relation Between Science and Technology

Many advances in technology are science applications, which might lead you to think that technology is secondary to science, but actually, there is a *dual* relationship between the two.

Quantum mechanics led to the invention of the *laser*, and the laser affected both science and technology. In science, the laser is an essential tool in many physics experiments; it even led to the emergence of new subfields of physics, such as atomic spectroscopy, and holography. In technology, the laser is used in manufacturing many products, and is part of many appliances, such as laser printers and optical discs.

From this relation between science and technology, we conclude that to accelerate both, we should implement latest technology in scientific experiments and apply latest scientific discoveries to technology. For this to happen there should be better communication between scientists and engineers. This communication should not be through published papers only; scientists and engineers should work together to share and discuss their ideas, and collaborate on projects with mutual interest.

There is also great need for academia-industry collaboration. Universities can provide advanced laboratories and many talented graduate students, while industry can provide funding and market expertise. Why then cannot they collaborate and share the intellectual property?

11.3.5 University Labs Versus Industrial Labs

In 1883, while working on his light bulb, Edison observed the flow of electricity across a gap, in vacuum, from a hot filament to a metal wire. Since he saw no practical application, he did not pursue the subject further. This phenomenon became known as the "Edison effect", and if he did investigate it, he might have shared the Nobel Prize with Owen Richardson, who analyzed the behavior of electrons when heated in vacuum [9].

This anecdote illustrates a big difference between industrial and university research. In Industry, researchers usually work only on problems that have practical significance, and although this kind of research is important, the downside is that they might overlook something as Edison did, and once they solve a problem they might not have the time or interest to pursue it in a more generalized setting.

In university, by contrast, researchers have the freedom to pursue the research they like, which might not have immediate practical application, but have greater impact in the future. However, they do not have as much resources as in industrial labs, and it might take longer for their research to find applications in industry.

Obviously, both kinds of research is indispensable, but how can we enhance their role? Researchers from university and industry should collaborate on problems of common interest; they will get to work on different kinds of problems than they are used to, and benefit from each other's point of view.

11.3.6 Improving the Publishing Process

In October 2013, *Science* magazine published the results of a "sting operation" conducted on open access journals. They sent a spoof paper to 304 journals. The paper was too flawed to be publishable; yet, 60 % of the journals accepted it [10].

This shocking result illustrates that some publishers seek only profit from open access journals. "Beall's list of Predatory Publishers" [11] includes 477 such publishers, from which 137 journals were used in the *Science* study and 82 % accepted the paper. All researchers should report such journals and avoid submitting papers to them.

The *Science* study showed another observation: about 90 % of the journals that accepted the paper used either superficial peer review or no peer review at all. Peer review is very important to help authors improve their papers, and to exclude flawed ones. However, peer review was criticized [12, 13] for causing bias towards the views of referees, and for failing to spot some flawed papers. What can we do to improve the peer review process?

A great idea to improve peer review is to do it after publication. This is the idea behind the *F1000Research* journal; it publishes papers only after a cursory quality check, peer review happens after publishing by referees who post their names, authors then can post comments and revisions when needed [14].

In my opinion, journals should apply this idea, but make any researcher, who published a few papers on the topic, able to review. I also suggest making a site that collects metadata on all published papers and allows researchers to *rate* them based on quality and significance. I think rating papers can be very efficient in identifying good research. In addition, rating can become another measure for significance, besides citations.

11.3.7 Publishing Negative Results

Nowadays many researchers work independently on the same problem; it is normal that many of their approaches fail to produce positive results. It is important that researchers publish those negative (null) results to save their colleagues' time and allow them to pursue other approaches more likely to succeed.

Refraining from publishing negative results leads to *publication bias* [15]. Suppose you are investigating the effectiveness of a new treatment but your results did not exclude the null hypothesis (i.e. the treatment is not effective). Meanwhile, someone else did another study, found the treatment effective, and published the results. If you do not publish your negative result, you cause a biased impression about the effectiveness of the treatment. This does not apply only to medicine, but also to new devices and experimental techniques.

The problem is that most negative results do not get published; why is that? Many researchers are reluctant to publish negative results because they underestimate their importance, or lose interest in the topic. Further, most journals do not publish negative results; they prize original papers.

There are some journals that publish only negative results, such as the All Results Journals (Chemistry, Biology, Physics and Nanotechnology), and the Journal of Negative Results in BioMedicine. However, they get very few submissions.

The entire scientific community should respect and acknowledge negative results; they should cite them, and regard them as valuable pieces of information. Journals must accept negative results, and if they do not, researchers should boycott them. Publishing negative results is not an option, it a *duty*.

11.3.8 Reproducing Research Findings

Scientists at the biotechnology firm Amgen tried to confirm the results of 53 papers that were considered landmarks in cancer studies, but they succeeded in confirming only 6 papers (11 %) [16]. Of those 53 papers, 21 were published in journals with impact factor greater than 20. Furthermore, the mean number of citations of non-reproduced papers was 248. This appalling result is a clear example about the importance of *reproducing* research findings; imagine how many scientists wasted their time and resources pursuing false results.

The reasons for this might, in small part, be outright fraud, but mostly scientists might unconsciously ignore results that contradict their claim; also, in complicated experiments, small errors can make big difference in the final result. A detailed discussion for the reasons of false results is offered in [17].

To solve the problem of reproducibility, the *Reproducibility Initiative* was created in 2008. The idea is that every submitted experiment is sent to an appropriate Science Exchange lab to reproduce it; then, the paper gets a badge for being "Independently Validated", and the validation results get published in the PLOS reproducibility collection [18].

The problem with this idea is that not many researchers would pay, or find a grant, to get their results validated, and even if they did, how sure can we be that the reproduced results are more correct?

In my opinion, reproducibility should be the role of the industrial companies that use the research results, such as pharmaceutical companies. In addition, researchers should not take for granted the validity of research results; if they want to work on a follow-up of a particular result, they should reproduce it and publish the reproducibility result with their original research, and journals must not reject those reproduced results.

11.3.9 Managing Research Literature

In the past, research results took months, even years, to get published. Now, the Internet solved this problem through preprints, and researchers are facing the problem of too many papers. For example, only in March 2014 more than 8,000 papers were submitted to arXiv [19].

To keep up with the huge amount of literature, researchers use reference manager software, they make summaries and notes, and they rely on review papers to provide an overview of a particular topic. However, review papers are usually for topics in which many papers were written already, and they can get outdated quickly.

I suggest making review papers like *wikipedia* articles; once a few papers appear on a new topic, a wiki-review paper is written summarizing their results and suggesting future research. Then, when another paper is written on the subject, its author, or someone else, summarizes it in the wiki-review. This goes on until we have a complete review on the topic.

11.3.10 Encouraging Multiple Research Approaches

In 1916, there was some evidence for the existence of black holes, but many scientists, most prominently Einstein and Eddington, refused to believe in their existence. In the 1930s, Chandrasekhar proved that stars heavier than 1.4 the mass of the sun cannot become white dwarfs, but Eddington and others gave *incorrect* arguments against the idea. Black holes were taken seriously in the 1960s when evidence became hard to ignore; however, much of the research done in the 1960s could have been done *fifty* years earlier [20, p. 138].

A similar situation happened in geology. In 1912, Alfred Wegener proposed the continental drift theory, but it was met by opposition from the majority of scientists, until the idea of plate tectonics was proposed in 1958 [21].

History of science teaches us not to hold to unjustified assumptions, even if they are held by the majority, we should always consider opposing views. Unfortunately, we are making the same mistake again; currently the vast majority of physicists working on approaches to quantum gravity and unification are string theorists. String theory is not the only approach to quantum gravity, but the other approaches receive little attention.

Concentrating on one approach to a given problem is not healthy for the progress of science, and the way out is offered by Lee Smolin in his book "The trouble with physics"; his solution can be summarized in the following points [22, pp. 351–353]:

1. Departments should ensure that rival research programs are represented on their faculties.
2. Scientists should be hired and promoted based only on their ability, disregarding their research approach.
3. All scientists should keep an open attitude towards competing research approaches, and encourage their students to learn about them.
4. Funding agencies and foundations should support scientists with different views from the mainstream.
5. A research program, or a particular view, should not be allowed to dominate any field without convincing scientific proof.

11.3.11 Encouraging Innovation in Global Problems

Currently, environmental and sustainability problems are among the biggest problems faced by humanity. I believe the key to solve those problems is by encouraging scientific and technological innovation.

In the middle of the twentieth century, a series of scientific and technological innovations in agriculture helped increase worldwide food production, which saved over a billion people from starvation; this is known as the "Green Revolution".

A promising innovation in energy is the Traveling Wave Reactor under development by TerraPower. This reactor has high fuel utilization rate, and does not need uranium enrichment or resupply. The reactor can be buried underground and run for 100 years [23].

These examples show that science has the potential of solving our problems, but it needs our support. Research and development does not receive enough funding. For example, in 2013 the US spent only $2 billion on clean energy R&D, compared with $72 billion on defense R&D [24]. Governments should create research organizations that join scientists and engineers to work together to find innovative solutions and implement them.

11.3.12 Funding Research and Development

In 2011, The United States spent $424 billion on research and development (R&D), which represented 30 % of the estimated $1.435 trillion spent globally on R&D. The EU spent 22 %, China 15 %, and Japan 10 % [25].

In 2001, the number of research papers from China represented 3 % of the world total. In 2011, China's share increased to 11 %, becoming the world's second-largest producer of scientific articles, after the US. This correlated with an increase in funding R&D from 1.0 % of GDP to 1.8 %. This correlation between funding R&D and research output appears in all the Newly Industrialized Countries (NICs).

All countries should understand this relation between funding research and development; money spent on research is an investment that is key for better future.

11.3.13 The Role of Scientists in Global Decisions

Global problems take a long time to be solved, especially environmental ones. They require coordination between nations for decades. Yet, politicians take those global decisions. Politics is local, and politicians usually care only about their election period. Scientists, on the other hand, study these problems for years, and their judgment is only affected by observation and experiment. Why then is the role of scientists limited to advising policy makers?

Scientists should have a more active role in global issues. Parliaments should form committees, mostly of scientists from various disciplines, to prepare the country's policies towards environmental, and other global, issues. These committees' decisions should be considered as a *requirement* for the government to follow, not a suggestion.

Scientists should also exert greater effort in raising the public awareness of environmental, and other global, problems. In 2011, *Eurobarometer* found that for 95 % of Europeans, protecting the environment was personally 'very important' or 'important'. Whereas, in the US a similar survey found only 63 % of Americans share that opinion [26].

The public opinion on environmental issues is reflected on their countries' environmental policy. In the EU, in 2012, 14.4 % of energy was from renewable sources, and the target is 20 % by 2020 [27]. In the US, only 11 % of energy production was from renewable sources [28], and there is no mandatory national target in 2020.

11.3.14 Raising the Public Understanding of Science

"Does the Earth go around the Sun, or does the Sun go around the Earth?" In 2012, a sample of US residents were asked that question, and 26 % answered incorrectly. In the EU a similar question was asked, in 2005, and 34 % answered incorrectly [29].

In 1990, when a sample of US residents were asked about their assessment of astrology, 35 % said it was 'sort of scientific' or 'very scientific'; that number increased to 42 % in 2012 [29].

Why do many people not know such basic scientific facts? Why do they believe in pseudoscience? Because education needs to improve and not enough is being done to raise the public understanding of science.

In democratic political systems, the public decides how much funding goes to science, and how scientific discoveries should be applied. When they understand more about science, they will be able to make better decisions on scientific issues, such as the environment, biotechnology, medical trials, and nutrition. They will also be less prone to be deceived by pseudoscience.

Most of the efforts done to raise the public understanding of science are focused on popular books, magazines, science fairs, and museums; these methods are clearly ineffective. When a sample of US citizens were asked about their primary source of information about science, they answered: the Internet 42 %, the TV 32 %, newspapers and magazines 15 %, books 3 % [30]. Clearly, the Internet and TV have the greatest influence on the public attitude and knowledge about science.

Communicating science to the public should not be the responsibility of individual scientists, but the entire scientific community should participate in an organized effort to do so. Efforts through the Internet should not focus only on scientific news, but also on established scientific facts in short articles containing images and videos. In addition, more TV documentary series should be made to discuss science in an interesting way.

11.3.15 Improving Education

Humanity needs scientists and engineers to accelerate progress. Inspiring a passion for science and technology in students is the first step towards that goal. Education, however, needs urgent improvements, and much has been written about that. That is why I will not discuss pre-university education, but talk about undergraduate education.

Universities are the place where scientists and engineers are trained, but most universities are not giving enough effort to education. Carl Wieman, who is a Nobel winning physicist, says [31] "There is an entire industry devoted to measuring how important my research is... Yet, we don't even collect data on how I am teaching."

Wieman developed a new method for teaching undergraduates based on active learning and deliberate practice. To test his method, an introductory physics class at the University of British Columbia (UBC) was split in two groups, one used Wieman's method and the other used traditional lectures. The first group scored twice as high as the control group on a multiple choice test. Wieman's method is now widely used at UBC [32].

To improve undergraduate education, Wieman suggests that universities should release data on their teaching methods as a condition for federal funding. Students then can use these data to decide which university to choose, thus forcing universities to improve their teaching methods [31].

11.3.16 Empowering All Humanity to Participate

In 2008, the total number of research papers from all African countries (55 countries) was about 27,000 papers [33]. For comparison, the US produces more than 310,000 papers annually [34]. Africa contains 1.1 billion people, three times those in the US; if that number effectively contributed to building the future, imagine how much difference they would make.

Many problems need remedies in Africa, but in my opinion, *education* is the most important factor in the development of any country. The African Institute for Mathematical Sciences (AIMS) is a fine example for the efforts that need to be done in that direction. AIMS was established in 2003 in South Africa, and 479 students from 34 countries have graduated from it [35]. Organizations working on development in Africa should consider supporting and establishing similar educational institutes.

However, to spread high quality education to many people, use the Internet. In the past few years, Massive Open Online Courses (MOOCs) witnessed a big increase in the number of students and courses. Coursera, the biggest MOOC platform, started in April 2012, and by January 2014, it had over 22 million enrollments from 190 countries across 571 courses [36].

MOOCs only need a computer and an Internet connection, but many African countries have bad Internet connection, and not many computers. However, some measures can easily be done to solve these problems:

- The One Laptop per Child project [37] is a great example on how to provide students with low cost computers.
- Local schools and universities should open their computer facilities to all learners.
- Developing organizations should create computer facilities in major cities for online learning.
- MOOCs should include pre-university courses.
- MOOCs should make verified certificates free, at least for learners from developing countries.

11.4 Conclusions

I believe science will be our guide to the future, but we need to improve the way we do science to accelerate the rate of scientific discovery and its applications. This is crucial to find urgent solutions to humanity's problems and improve humanity's conditions. In this essay, I have tried to identify those aspects of science that need improvement, and discuss how to improve them. Those issues are complex and no one essay can adequately cover them. We need the entire scientific community to pay attention to those issues. I hope this essay will open a window for discussion in that direction.

References

1. Hawking, S.: The Universe in a Nutshell. Transworld Publishers, London (2001)
2. Feynman, R.P.: There's plenty of room at the bottom. Eng. Sci. **23**(5), 22–36 (1960)
3. Ron, J.M.S.: The past is prologue: the future and the history of science. In: González, F. (ed.) There's a Future: Visions for a Better World. T.F. Editores, S.L.C. (2012)
4. Drexler, K.E.: Engines of Creation: Challenges and Choices of the Last Technological Revolution. Anchor Press Garden City, New York (1986)
5. Moss, F.: The power of creative freedom: lessons from the MIT Media Lab. In: González, F. (ed.) Innovation: Perspectives for the 21st Century. T.F. Editores, S.L.C. (2011)
6. The University of Science and Technology: Curriculum structure
7. Appell, D.: The supercollider that never was. Scientific American. http://www.scientificamerican.com/article/the-supercollider-that-never-was/. Accessed 28 Jan 2015
8. International Space Station: How much does it cost? http://www.esa.int/Our_Activities/Human_Spaceflight/International_Space_Station/How_much_does_it_cost. Accessed 28 Jan 2015
9. Rosenberg, N.: Innovation: it is generally agreed that science shapes technology, but is that the whole story? In: González, F. (ed.) Innovation: Perspectives for the 21st Century. T.F. Editores, S.L.C. (2011)
10. Bohannon, J.: Who's afraid of peer review? Science **342**(6154), 60–65 (2013)
11. Beall, J.: List of predatory publishers. http://scholarlyoa.com/2014/01/02/list-of-predatory-publishers-2014/ (2014). Accessed 28 Jan 2015
12. Lee, K., Bero, L.: What authors, editors and reviewers should do to improve peer review. Nature **471**, 91–94 (2006)
13. McCook, A.: Is peer review broken? The Scientist **20**(2), 26 (2006)
14. Rabesandratana, T.: The seer of science publishing. Science **342**(6154), 66–67 (2013)
15. Dirnagl, U., et al.: Fighting publication bias: introducing the negative results section. J. Cereb. Blood Flow Metab. **30**(7), 1263 (2010)
16. Begley, C.G., Ellis, L.M.: Drug development: raise standards for preclinical cancer research. Nature **483**(7391), 531–533 (2012)
17. Ioannidis, J.P.A.: Why most published research findings are false. PLoS Med. **2**(8), e124 (2005)
18. The Reproducibility Initiative. http://validation.scienceexchange.com/#/reproducibility-initiative. Accessed 28 Jan 2015
19. ArXiv monthly submission rates. http://arxiv.org/stats/monthly_submissions. Accessed 28 Jan 2015
20. Thorne, K.S.: Black Holes and Time Warps: Einstein's Outrageous Legacy. W.W. Norton & Company, New York (1994)
21. Oreskes, N.: From continental drift to plate tectonics. Plate Tectonics: An Insider's History of the Modern Theory of the Earth. Westview Press, Boulder (2001)
22. Smolin, L.: The Trouble with Physics. Mariner Books, New York (2007)
23. LLC TerraPower: Traveling-wave reactors: a truly sustainable and full-scale resource for global energy needs. In: Proceeding of ICAPP 2010 (2010)
24. National Science Foundation: Federal R&D funding by budget function: fiscal years. http://www.nsf.gov/statistics/nsf14301/pdf/nsf14301.pdf (2012–14). Accessed 28 Jan 2015
25. National Science Foundation: Research and development: national trends and international comparisons. http://www.nsf.gov/statistics/seind14/index.cfm/chapter-4/c4c.htm. Accessed 28 Jan 2015
26. National Science Board (US): Science and engineering indicators. http://www.nsf.gov/statistics/seind14/index.cfm/chapter-7/c7s4.htm (2014). Accessed 28 Jan 2015
27. EurObserv'ER: Estimates of the renewable energy share in gross final energy consumption for the year. http://www.eurobserv-er.org/pdf/press/year_2013/res/english.pdf (2012). Accessed 28 Jan 2015
28. U.S. Energy Information Administration: Monthly energy review. http://www.eia.gov/totalenergy/data/monthly/. Accessed 28 Jan 2015

29. National Science Board (US): Science and engineering indicators. http://www.nsf.gov/statistics/seind14/index.cfm/chapter-7/c7s2.htm (2014). Accessed 28 Jan 2015
30. National Science Board (US): Science and engineering indicators. http://www.nsf.gov/statistics/seind14/index.cfm/chapter-7/c7s1.htm#s3 (2014). Accessed 28 Jan 2015
31. Mervis, J.: Transformation is possible if a university really cares. Science **340**(6130), 292–296 (2013)
32. Chasteen, S.V., Perkins, K.K., Beale, P.D., Pollock, S.J., Wieman, C.E.: A thoughtful approach to instruction: course transformation for the rest of us. J. Coll. Sci. Teach. **40**(4), 70–76 (2011)
33. Adams, J., King, C., Hook, D.: Global research report Africa. Evidence, a Thomson Reuters business (2010)
34. Adams, J., Pendlebury, D.: Global research report United States. Evidence, a Thomson Reuters business (2010)
35. About AIMS. http://www.aims.ac.za/en/about/about-aims. Accessed 28 Jan 2015
36. Coursera: Our student numbers. https://www.coursera.org/about/community. Accessed 28 Jan 2015
37. OLPC's mission. http://one.laptop.org/about/mission. Accessed 28 Jan 2015

Chapter 12
How to Avoid Steering Blindly: The Case for a Robust Repository of Human Knowledge

Jens C. Niemeyer

Abstract Steering the future hinges on the availability of scientific and cultural data from the past. As humanity transitions into the digital age, global access to a condensed form of human knowledge becomes a realistic technological possibility and potentially a human right. At the same time, the risk of losing the vast majority of this information after a global disaster has never been greater. I argue that a collaborative effort to create a secure repository of human knowledge would not only protect humanity's cultural heritage for future generations, it could also define a minimum standard for the information that every human being should have a right to access. The basic requirements and challenges for creating the repository are discussed.

12.1 Introduction

Throughout human history, two elements have proved essential for the development of human culture: the capability to conserve and communicate previously gathered knowledge about humans, their interactions, and the world they live in, and the capacity to derive "useful" (in the sense of being beneficial for the long-term survival and development of the respective group of humans) decisions based on the available information. The basic assumption of this essay is that this observation will remain true for the way humanity will steer the future.

The second element, decision making, has been the responsibility of tribal leaders, monarchs, or elected officials using guidance provided by different types of expert committees (elders, spiritual advisors, scientists etc.) which, in turn, have used the best available sources and technologies at their disposal. Looking forward, great efforts are being made to employ simulations of complex social and economical interactions to advance the technical support for decision-making (e.g., the "Living Earth Simulator" [1]). While this is undoubtedly a central ingredient in humanity's capacity to steer the future, it will require major technological breakthroughs and its

J.C. Niemeyer (✉)
Institut für Astrophysik, Universität Göttingen, Göttingen, Germany
e-mail: jens.niemeyer@phys.uni-goettingen.de

© Springer International Publishing Switzerland 2016
A. Aguirre et al. (eds.), *How Should Humanity Steer the Future?*,
The Frontiers Collection, DOI 10.1007/978-3-319-20717-9_12

ultimate success may be fundamentally constrained by limitations in the predictiveness of models of complex systems far from equilibrium.

By comparison, the first element, i.e. conserving and making accessible the essential parts of human knowledge (where defining the metric for "essential" will be one of the key challenges) into the far future, has received much less attention. This is critical, because as humanity transitions into an age where information exists nearly exclusively in digital form, this knowledge is far more vulnerable to major disruptions of technological infrastructure than ever before. The risk of losing most our of cultural and scientific heritage after a global disaster (for instance, any event that interrupts electrical power supply on a global scale for more than a couple of weeks) is substantial unless precautions against this *digital amnesia* are taken.

On the other hand, digital technology also offers the opportunity to create a universally accessible, multi-dimensional and multi-resolution repository of human knowledge (henceforth simply referred to as the *repository*) with an enormous potential for research and education in "normal" times. As will be argued below, the internet already provides several promising services for information conservation and retrieval, but none of these satisfy all of the central requirements of the envisioned repository. No fundamental breakthroughs are needed to begin this project, and the risk of digital amnesia adds a certain sense of urgency. The goal of this essay therefore is to demonstrate the potential benefits of a global knowledge repository (Sect. 12.2), to summarize the basic requirements derived from these intended purposes (Sect. 12.3), and to outline some ideas for its creation (Sect. 12.4).

12.2 Benefits of the Repository on Different Time Scales

The diverse benefits of a global knowledge repository can be demonstrated most clearly by viewing the future of humanity on different time scales. This will allow us to identify its key requirements specifications.

12.2.1 Near Future

Already at the present, research and education profit greatly from open online resources that cover different parts of the spectrum of human knowledge. Wikipedia is probably the most prominent example for a very broad but relatively shallow organization of knowledge (we might call it "horizontal"), allowing the efficient retrieval of information about an extremely wide range of topics, but being poorly suited for an in-depth study of any given subject starting from zero and going to the frontiers of current research ("vertical" organization). On the other hand, a wide variety of courses and tutorials ranging from short How-To clips on YouTube to postgraduate level Massively Open Online Courses (*MOOCs*) are available for the latter.

Ideally, a knowledge repository should allow flexible navigation in depth and breadth at an arbitrary level of detail ("multi-resolution") within the space of (appropriately linked) fields of knowledge ("multi-dimensional"). In other words, it should provide overlapping maps of the space of human knowledge with adjustable resolution, i.e. an atlas of knowledge space. Clearly, the atlas must continually evolve by ingesting new information and re-organizing existing correlations. This can only be achieved if the repository acquires a certain (and growing) degree of *autonomy* using artificial intelligence, combined with human supervision, for creating and updating maps.

The closest existing structure to the proposed repository is perhaps the *Internet Archive* [2] operated by a non-profit organization that aims to provide permanent access to historical collections in digital format. However, the repository's purpose would go far beyond static preservation of data. In order to allow humanity to "bootstrap" its knowledge in the aftermath of a global disaster, great emphasis must be placed on i) the self-containedness and self-consistency of the information in the spirit of an enormous collection of cross-linked textbooks (starting with tutorials to read and write), and ii) the robustness of the data in the event of long-term disruptions of infrastructure (see below).

12.2.2 Intermediate Future

On the scale of several decades, free access to essential knowledge must (and probably will) become a major political goal and may be considered a basic human right. The definition of "essential knowledge" obviously depends on the cultural and religious perspective. Nevertheless, it is crucial for the future of humanity that a consensus can be found on what information is considered essential for conservation into the far future, where "information" refers to facts and all the necessary correlations to understand them from scratch (the *maps* introduced above). At the same time, this body of knowledge can define the minimum level of information that every human being should be entitled to having access to. We should hence add *accessibility* to our list of key properties of the repository.

If the repository project becomes sufficiently visible, it can perhaps trigger a global political and cultural debate about the importance of access to information as a human right. This may indeed contribute substantially to help humanity steer the future.

12.2.3 Far Future

Looking into the more distant future, humanity will face global disasters (cometary collisions, epidemics, nuclear wars etc.) with finite probability. Some of these will disrupt human society and its technological infrastructure for long periods of time.

One of the main objectives of the knowledge repository must be the conservation of humanity's cultural and scientific heritage and the "re-booting" of human society after such catastrophic events.

This purpose of the repository is similar in spirit to the mission of the "Albertian Order of Leibowitz" in Walter M. Miller's classic novel [3].[1] However, my vision of the future and of humanity's handling of scientific knowledge is considerably more positive than Miller's.

The *Long Now Foundation* [4] is one of the few institutions that support long-term projects on these timescales. Their *Manual for Civilization* shares some of the goals of the proposed repository.

Currently, no widespread efforts are being made to protect digital resources against global disasters and to establish the means and procedures for extracting safeguarded digital information without an existing technological infrastructure. Facilities like, for instance, the *Barbarastollen* underground archive for the preservation of Germany's cultural heritage (or other national and international high-security archives) operate on the basis of microfilm stored at constant temperature and low humidity. New, digital information will most likely never exist in printed form and thus cannot be archived with these techniques even in principle.

The repository must therefore not only be *robust* against man-made or natural disasters, it must also provide the means for accessing and copying digital data without computers, data connections, or even electricity.

Speculating about the ultra-distant future, mankind may spread out inside the solar system or beyond. In this case, the repository would fulfill yet another purpose: the condensed human knowledge to be carried along wherever humans travel.

12.3 Fundamental Requirements

We can now address each requirement in turn to discuss the main challenges for the development of the repository.

12.3.1 Semi-autonomy

Long-term data archiving is already an active field of research in the information sciences. However, the envisioned knowledge repository is far more ambitious than any existing approaches since it intends to organize *all* forms of knowledge into an atlas of overlapping maps. Furthermore, none of this information can actually be archived into slow-access storage, as the continuous map updating process will require fast access to essentially all of the data.

---

[1] An alternative title for this essay might be "What would Leibowitz do in the age of ebooks?".

The human brain is the most powerful known realization of a flexible storage and retrieval mechanism for multi-dimensional and multi-resolution facts and correlations. Creating and updating maps from a continuous data flow is similar to committing memories to long-term memory in the human brain. It is therefore plausible to assume that brain research, such as the Human Brain Projects funded prominently in Europe and the US [5], may provide useful insights for the software design of the repository. More generally, further research in neurophysiology and artificial intelligence are needed to make the map-creating process partly autonomous in the long run to keep up with the ever-increasing amount of data.

While semi-autonomy of the repository for data ingestion and re-organization will be crucial on longer time scales, it is important to stress that it is not a pre-requisite for starting the project. Much of the initial structure of the repository can and must be constructed by human developers.

12.3.2 Accessibility

During times of a fully operational global technological infrastructure, access to digital information from nearly every place on the planet presents only practical, but no fundamental technological obstacles. The biggest challenge here might be the avoidance of a pronounced social gradient, i.e. access to digital information must be provided for all humans regardless of their social, geographical, or political status. A noteworthy example of a project with similar goals is the *Outernet* initiative [6].

The repository project can immediately aim for a free, multi-device internet service. It would be desirable to keep it independent from corporate and political interests as far as possible.

In contrast, providing accessibility in the event of a global disaster presents a serious technological (and sociological) challenge for the project. The repository must feature a "bootstrapping mode" which allows the regrowth of knowledge from nearly zero. A few ideas will be sketched below, but clearly this question warrants a dedicated research initiative.

12.3.3 Robustness

Despite some efforts to build ultra-secure data storage facilities, it is probably safe to assume that most existing data centers would not survive a disaster on continental or global scales for more than several months. Stability of power supply and finite redundancy against hardware failure are among the most serious long-term problems.

On the other hand, not all of the data would have to be directly accessible for all times. Instead, the repository could allow for a hierarchy of resolution and dimensionality of the stored knowledge maps in a sort of onion-skin structure, as illustrated in Fig. 12.1. The outermost layer of data, having the highest level of redundancy and

Fig. 12.1 Illustration of the onion-skin strategy for protecting the repository against global disasters. The outer layers should provide robust and redundant access to basic information (tutorials and instructions) which is then used to successively recover the inner layers

post-crisis accessibility, would only need to contain the information needed to restore technology to the level where the next layer can be resurrected, and so on until all the necessary infrastructure has been recreated to access the entire data (which might be centuries later).

For example, a strategy that combines ultra-secure, centralized data centers with massively distributed independent devices (along the lines of the *One Laptop per Child* project [7], albeit with different goals and distribution strategies) might guarantee the intended degree of robustness and redundancy. In this case, the independent devices (e.g., solar powered, low-cost tablets) would represent the outermost layer of the onion skin, containing, for instance, simple tutorials for survival, basic social and technical infrastructure, as well as instructions on how to find and operate the next level storage facilities, which might be local data centers, etc. At the core of the onion, underground facilities designed to survive for centuries without maintenance (such as, e.g., [8]) would store the highest resolution knowledge maps. Thinking very far ahead, even a lunar outpost of the repository might be considered to be on the safe side.

12.4 Outlook: How to Get There

First steps toward the implementation of the repository can be taken immediately. In the beginning, a crowdsourcing approach might be most promising, supported by a coordinated initiative to attract research institutions to work on the fundamental design challenges regarding robustness and semi-autonomy described above. The goal of this initial phase would be the establishment of a global "repository collaboration" and first demonstrations of the feasibility of the project.

In the long run, however, the repository should strive for some sort of global political recognition that supports (if only symbolically) its relevance for the future of humanity and protects its existence and independence. This might be achieved, for instance, under the umbrella of UNESCO.

Ultimately, the protection and support of the repository may become one of humanity's most unifying goals. After all, our collective memory of all things discovered or created by mankind, of our stories, songs and ideas, have a great part in defining what it means to be human. We must begin to protect this heritage and guarantee that future generations have access to the information they need to steer the future with open eyes.

Acknowledgments I would like to thank the FQXi and its sponsors for their support of this essay contest, and the participants of the discussion forum for many interesting comments.

References

1. Paolucci, M., Kossman, D., Conte, R., Lukowicz, P., Argyrakis, P., Blandford, A., Bonelli, G., Anderson, S., Freitas, S., Edmonds, B., et al.: Eur. Phys. J. Spec. Top. **214**(1), 77 (2012). arXiv:1304.1903
2. https://archive.org
3. Miller, Jr., W.M.: A Canticle for Leibowitz. Bantam Spectra, New York (1984). ISBN 9780553273816
4. http://longnow.org
5. Kandel, E.R., Markram, H., Matthews, P.M., Yuste, R., Koch, C.: Nat. Rev. Neurosci. **14**(9), 659 (2013)
6. https://www.outernet.is
7. http://one.laptop.org
8. http://www.swissfortknox.com
9. An alternative title for this essay might be "What would Leibowitz do in the age of ebooks?"

Chapter 13
The Tip of the Spear

George Gantz

> The day will come when, after harnessing space, the winds, the
> waves, the tides and gravity, we shall harness for God the
> energies of love. And on that day, for the second time in the
> history of the world, man will have discovered fire [1].
> > Pierre Teilhard de Chardin.

Abstract The evidence is clear—there is a new emergent phenomenon arising from the global integration of human knowledge and aspirations linked through advanced networks. As in each previous emergence of higher order from lower, the behaviors that evolve from the complex interaction of the individual components cannot be predicted. Can we influence the trajectory of this emergence in ways that benefit the individuals that comprise it and increase the probabilities of continued progress? In addition, can we prepare for the potentially rare but nevertheless real possibility of first contact with an extraterrestrial civilization? Yes, by drawing on evolutionary lessons to identify and promote collectively beneficial behaviors in our global institutions, including the institution of science. As human civilization continues to evolve, progress will be powered by knowledge, but we should arm "the tip of the spear" with the human empathic values of trust, humility, mutual respect and shared commitment: in a word, with love, in its most universal form.

13.1 Introduction

Human civilization, it may be argued, began when humans learned to control fire [2]. The technology of fire gave early humans a level of mastery and control over their environment that enabled subsequent biological, cultural, economic and technological developments. Humanity flourished, extending its dominion across the globe. The exclusive authority of environmental factors to shape the future ceded to human influences. The Pleistocene ended and the Anthropocene began.

G. Gantz (✉)
83 Claypit Hill Rd, Wayland, MA 01778, USA
e-mail: gantz1@mac.com

© Springer International Publishing Switzerland 2016
A. Aguirre et al. (eds.), *How Should Humanity Steer the Future?*,
The Frontiers Collection, DOI 10.1007/978-3-319-20717-9_13

We are again facing a massive global transition driven by the influence of human networks. Human technology and engineering has enabled increasingly complex networks to develop among the now eight billion humans on earth. Individual human behaviors are now subsumed within a complex interplay of institutions (networks of humans)—the resulting dynamics of which drive global outcomes. The global civilization that is emerging from this new evolutionary process, operating at the institutional level, will exhibit behaviors that we may not be able to understand, predict or control. The power of human agents to shape the future may be ceded to global institutions that have evolved beyond our ability to manage. The Anthropocene may end soon after it began.

The rapid acceleration of technological change in the last century has allowed us to penetrate into space and to explore the very largest and the very smallest structures in the universe, while vastly improving the quality of life for most humans [3]. But there have been negative consequences as well, including human exploitation, institutional failures and unanticipated consequences, all facilitated by increasingly potent technology. There continue to be serious concerns that such consequences could include the extinction of the human race. Indeed, according to Sir Martin Rees, "I think the odds are no better than fifty-fifty that our present civilization on Earth will survive to the end of the present century." [4]. Some might say that humans, having evolved in primeval forests and savannahs, may not be up to the challenge of managing modern technology.

Evidence also suggests that technology has released the human race from the constraints of evolution by natural selection. Certainly, selection pressures applicable to human reproduction have changed—technology has significantly altered the human fitness landscape. In 2007, Freeman Dyson speculated that human cultural evolution replaced biological evolution about 10,000 years ago, and he further noted, "in the last 30 years, Homo sapiens has revived the ancient pre-Darwinian practice of horizontal gene transfer… blurring the boundaries between species." [5]. We may be moving into an era when human reproduction and genetics will largely be functions of personal preferences amidst shifting cultural norms, economic incentives and technological capabilities—an entirely novel set of selection pressures.

In this context of accelerating institutional complexities, increasing technological threat and the re-writing of human evolutionary dynamics, steering the future of humanity is a considerable challenge.

13.2 Emergence

Over the last century, observational science, mathematical theory and computational capabilities made significant advances that have opened up our understanding of complex systems and their emergence from the behaviors of individual component units [6, 7]. One key revelation is that our universe, including life itself, has evolved through a series of successive states, from low entropic, homogeneous conditions at the Big Bang, through increasingly complex states of higher entropy. The transition to each subsequent state involves a loss of symmetry, an increase in complexity and the emergence of novel structures and behaviors.

The process by which new structures emerge at each stage of the process is not uniformly well understood. Theories regarding the phase changes early in the history of our universe, leading to the emergence of the fundamental physical forces and the particles comprising the Standard Model, have a strong consensus, although major theoretical problems remain [8, 9]. Similarly, the theory of evolution through natural selection has a strong consensus in the scientific community, but debates continue on some of the specifics [10].

In his theory of evolution by natural selection, Darwin hypothesized the first modern emergence theory. Genetic mutations are introduced in individuals within a species and subjected to environmental selection pressures influencing reproduction, e.g. they compete for reproductive success. If a mutation is advantageous the individual will have a higher likelihood of reproducing, resulting in the spread of the mutation. The end result is an adaptive change in the population. Over time, diverse new behaviors and structures, including new species, arise. Analogous evolutionary processes in economics [11] and cultural behavior [12] also demonstrate evolution through innovation, competition/selection and reproduction, resulting in adaptive changes in the respective populations.

Theoretical discussions are now taking place, under various labels such as universal, quantum or cosmic Darwinism, speculating that each stage of emergence after the primordial Big Bang exhibits a kind of mutation and selection, with the emergent solution settling on "attractors" or "pointer states" with stable properties. The structures that survive this evolutionary process in a given environment are the ones that are optimally suited to the fitness landscape. One can visualize this process in the behavior of fluid flowing down a drain. Opening the drain initiates a flow that creates turbulence, during which a series of small structures may form spontaneously and be tested for fitness, quickly evolving to the efficient vorticular flow with which we are all familiar.

A deep insight is that attractors are formed in a process where the individual component units of the system, while behaving autonomously, are influenced by signals from other units. This results in changes, or mutations, in the local state of the system, which are then subject to selection pressures by the fitness landscape. The signaling and response between individual units is the basis for the self-organizing feature of the emergent process, and it is fundamentally a cooperative behavior. Innovations that exhibit greater cooperation among the units, for example by providing greater efficiency or stability, will out-perform those that do not. Debates continue on the degree to which such cooperative behavior exists in some, or all, emergent processes, and the extent to which it is consistent with reductionism or requires some form of top-down causation. In any event, the practical implications are clear. Successive emergent states are formed, in many if not all cases, through mutual interactions between component units of a system.

13.3 Cooperation

One of the historic criticisms of evolutionary theory is that it could not account adequately for the development of empathy and other moral qualities in human beings. After all, it seems counter-intuitive to suggest that a theory, colloquially referred to as "survival of the fittest," would result in cooperative rather than exclusively competitive behaviors. Recent research seems to have largely resolved these criticisms through models of multi-level individual and group selection processes [13] that demonstrate the evolutionary value of cooperative behaviors. Researchers have also suggested that evolution can account for the development of human morality [14] and human religions [15]. It no longer seems far-fetched to suggest that the higher moral and aspirational qualities of humanity have roots in the evolutionary heritage of our species. Moreover, the evolution of consensual moral frameworks and cooperative enterprise grounded in human empathy has been critically instrumental in our adaptation and subsequent success as a species.

Human adaption and advancement also required increasingly sophisticated forms of cooperation. As hunter-gatherer tribes were replaced by settled communities and the division of labor increased, the size and complexity of human networks increased. These networks became institutions as the underlying cooperative practices and behaviors were formalized. Governments, religions, markets, cultural and educational practices and organizations developed and evolved. Competition and innovation, within the landscape of the collective needs and aspirations of human individuals and groups, shaped the evolution of these institutions. Those bringing greater success in the accumulation of resources and the satisfaction of wants flourished and grew.

Among the successful institutional threads was the enterprise of natural philosophy. Greater empirical understanding of the world in which humans lived yielded significant benefits, and those who acquired and articulated such knowledge were highly valued, as were the libraries in which such knowledge was contained. In recent centuries, empirical science, the outgrowth of natural philosophy, has been the engine powering the technological change that has brought us to our present state. Through the cooperative efforts of scientists from all parts of the globe, knowledge of the world has increased and technology has flourished.

At the same time, many of our other institutions have also evolved, growing in size and sophistication, enabled by new technologies for communication, trade, travel, computation and manufacturing. This growth has brought profound benefits to the human species, but has also increased complexity and uncertainty. This complexity has given rise to novel behaviors, demonstrating emergence of higher-level structures [16]. These behaviors are not necessarily benevolent. According to Nassim Taleb, "…the world in which we live has an increasing number of feedback loops, … thus generating snowballs and arbitrary and unpredictable planet-wide winner-take-all effects" [17]. The daily news is headlined by the equally unpredictable behaviors of weather, stock markets and politics. The first is a complex phenomenon of nature, albeit increasingly influenced by human behavior. The other two are complex human institutional phenomena.

Cooperative enterprise is a hallmark of humanity's success. Humans consistently demonstrate trust in fellow humans, enabling our species to solve the Prisoner's dilemma, a game theory scenario that pits a rational betrayal against a more risky decision involving trust—if reciprocated, trust leads to a maximally beneficial outcome. Moreover, the institutions of human civilization all arose as networks of cooperating (or at least compliant) individuals. This cooperation provided the institutional foundation for the building of cathedrals, castles, commerce, computers and super-colliders. However, if the human evolutionary process that imbued humans with trust and facilitated the development of consensual moral frameworks has now been dismantled, how do we insure the continued selection and reinforcement of these qualities?

It is no small concern that the formative selection pressures of the fitness landscape that produced humans with immense cognitive strengths and powerfully cooperative behaviors may no longer be operating. Increasingly, we are faced with the challenge and responsibility for shaping humanity's future through intentional human design. We must create fitness landscapes that select for cooperative individual and institutional behaviors. Do we have the technical tools, the creative ideas and, most importantly, the collective will to do so?

13.4 Confrontation

Humanity has breached the earth's atmospheric barrier, first with man-made patterns in electromagnetic frequencies and later with exploratory artifacts and even vehicles. While this is a spectacular technical achievement, it also presages another concern for humanity. Has life evolved elsewhere? If it has, what will happen when we make contact?

Conventional wisdom had been that humanity is unique in the universe, and to date all efforts to detect evidence of extraterrestrial civilizations have been fruitless [18]. However, NASA has reported the discovery of organic materials on Mars, [19] and organic materials appear to be common throughout the universe [20]. New estimates of the number of potentially habitable planets in the Milky Way galaxy and the universe at large also suggest a much higher probability that life may have developed on other planets than previously thought [21].

It is conceivable that we will soon confront one, and potentially many, intelligent species from elsewhere in the universe. In this event, the future of humanity will depend on its ability to negotiate within a new, galactic-level fitness landscape. It would seem reasonable to expect that any sentient civilization with the technology and institutional capacity for space exploration will have completed an evolutionary process on their home planet that likewise solved the Prisoner's Dilemma through trusting behaviors and shared moral frameworks. The nature of that extraterrestrial morality and the cooperative behaviors it inspires may, however, be quite exotic.

How will our global institutions respond to first contact? Will political, military and scientific institutions cooperate in offering a united response, or will fear and confusion predominate? Will we be able to communicate our shared moral framework and negotiate a mutually beneficial outcome, or will technological supremacy determine a victor, resulting in horrendous costs?

This possibility may seem hypothetical, but we ignore potential Black Swan events at our peril. The asteroid that caused the K/T (Cretaceous–Tertiary) Extinction was a low probability event, but when it struck the earth the consequences were cataclysmic. So might be first contact.

13.5 Steering the Future

Human civilization is being challenged from within by accelerating technological progress and complexity, and may be challenged from without by first contact with extraterrestrial life. Historically the human response to challenge was often violence—hoisting a spear or other weapon in combat or conquest. However, the spear has also served humanity for both hunting and defense. While recent military jargon may have trivialized the "tip of the spear" analogy, it may yet have some value in our consideration of humanity's global emergence and potential first contact. Indeed, it is appropriate to ask what powers the spear of human civilization towards its unknown future, and how should we arm the tip?

The driving force of humanity's remarkable advance from the Pleistocene to the Anthropocene, including the mastery of fire, was the collective and shared learning about the world and the adaptation of that knowledge to our needs and desires. The human species has a passion for knowing, derived from necessity and enabled by bodies and brains of immense complexity and sophistication. That passion has found its greatest outlet in the empirical scientific discoveries of recent centuries. Yet those discoveries would have remained unexplored or unexploited without a corresponding institutional framework supporting freedom of thought and expression, dissemination and critical review of ideas and market demonstration, development and deployment. Universities replaced palaces. Free states replaced city-states. Trade in goods and ideas became global. The scientific community became a network of professionals that shared common goals and methods and achieved profound knowledge of the physical world. The foundation for all these achievements is the human empathic qualities that enable such cooperation.

It is essential that our human civilization remain committed to the pursuit of empirical knowledge. This will continue to be the power behind the spear. However, this pursuit is fundamentally dependent on maintaining institutional behaviors that support global cooperation. Trust, honesty, openness to criticism and new ideas, mutual respect and a passionate commitment to empirical truth have been essential to science and those qualities remain critical for sustained cooperation to exist within the scientific community. But is the fitness landscape for the scientific enterprise today selecting for these behaviors? Are the rewards and disincentives, the signaling and feedback loops, the administration and enforcement mechanisms within the enterprise properly aligned to achieve maximally cooperative behaviors? Or is the landscape of increasing specialization and fragmentation and increasingly steep incentives for being novel and being first, tending to undermine both cooperation and, ultimately, progress? Is the global institutional framework within which science does its work appropriately sympathetic and collaborative? Or is politicization and polarization undermining efficiency and fraying the shared moral framework under which it operates?

It may be difficult to answer these questions. Nevertheless, we must answer them. Humanity is the first species to have worked its way out of the confines of the natural fitness landscape—and we have the capability to design our own. This offers new degrees of freedom, and also brings with it responsibility for the consequences. For example, if we design, or fail to reform, institutions that do not engage in pro-social cooperation and that practice or enable cheating or defection, thereby undermining trust, then we risk having such institutions outrun their rivals in a winner-take-all competition. All of human civilization to this point would be in jeopardy, and we would have no one to blame but ourselves. However, if we embrace the centrality of cooperation to our evolutionary success and infuse it into our design of the fitness landscapes that determine future institutional success or failure, then we can take control of the future.

As we address this challenge, we must recognize that humanity is multi-dimensional and our interests extend beyond the material to include aesthetic, cultural, civic and spiritual aspirations. Institutions have evolved in all these dimensions, and their qualities, as in the case of science, have been shaped by human relationships. Institutions reflecting and reinforcing empathic qualities, whether families, tribes, cities, kingdoms, nations, religions, social movements or voluntary associations, benefit from cooperative behaviors, build social capital, and tend to thrive. (For example, efficient global markets are impossible to achieve without trusting relationships [22].) Those that do not, such as despotic autocracies, carry within a weakness in human bonding that undermines flexibility, responsiveness and information flow, all of which are essential for long term institutional success in satisfying human needs and aspirations.

These institutions also form networks and interact with each other. The institution of science, for example, depends on supportive economic and political institutions, and it, in turn, influences civic and cultural life. Ultimately, human civilization is the totality of human institutions and their collective behavior. As in other complex systems, institutions signal and respond, and the resulting behaviors are tested in a global fitness environment. Cooperative responses create synergies that lead to efficiencies and improved fitness—and therefore institutions that reinforce empathic behaviors should be respected as part of the global institutional framework that has also enabled science. Competitive or conflicting responses create frictions that can undermine or destroy—such institutional conflicts should be subject to negative selection pressure.

The 20th century has clear examples of both collaboration and conflict. Autocratic government paired with communist ideology contributed to the rise of Stalinism. Parochial nationalism and secular idealism contributed to Nazism. Thankfully, both failed to achieve their goal of global conquest. However, the competitive conflicts of World War II and the Cold War that defeated them resulted in massive loss of human life and waste of global resources. On the other hand, some collaborative global institutions have flourished. Science is a largely borderless enterprise that accumulated sufficient civic and economic support to build, among other things, the Hubble Telescope, the Human Genome Project and the Large Hadron Collider. In addition, market economies have thrived as global cooperation expanded—the flow of goods and services has evolved into an unrecognizably complex web of materials, components and services that defy efforts to comprehend it [23]. The United Nations is an example of a nascent synergism that continues to be tested in a fitness landscape that includes global political and economic conflicts.

Science does not always serve in an empathic capacity. Nuclear armament, with its potential for causing human extinction, is a clear example. Less clear is the role science may play in fostering particular ideologies such as determinism and materialism, metaphysical worldviews that arguably challenge the efficacy of human empathy and undermine the emotional and psychological foundation of other key human institutions—including religions—that promote empathy. Has science as an institution contributed to existential alienation, the rise of unfettered commercialism or declines in social capital and shared moral frameworks?

It is clear that the qualities that propelled humanity and its institutions forward are the empathic qualities of trust, honesty, mutual respect and shared commitment. To this list we should add the corollary attribute of humility. As Francis Bacon put it more than four hundred years ago, referring to both science and religion, "let men endeavor an endless progress or proficiency in both; only let men beware that they apply both to charity, and not to swelling" [24].

Without these empathic qualities, the human race would never have advanced and likely would not have survived. Without them, it is unlikely that we will survive.

While the evolutionary theories cited in this essay may be new, the idea that empathy is the foundation of human civilization is not. Indeed, one formulation of the behavioral foundations for human cooperation was promulgated thousands of years ago, in the Decalogue [25]. Both the Buddhist and Christian traditions emphasize compassion and love, respectively. Christianity specifically commands, "Love your neighbor as yourself" [26].

The advance of our human enterprise will be powered by empirical knowledge, but the tip of the spear should be armed with our empathic qualities, ensuring that it is a tool of advancement and not destruction, a probe rather than a weapon. As a civilization we must aspire to practice empathy and to build empathic qualities into our institutions. We must design the fitness landscape for humanity's future in ways that reward cooperation and collaboration and discipline cheating, dishonesty and other moral defections—thereby reinforcing the qualities of trust, honesty, mutual respect, humility and shared commitment. In so doing we will ensure the success of our collective enterprise as a whole and an optimal outcome from interactions with civilizations we have yet to meet.

13.6 Conclusion

Human civilization is facing many challenges in the 21st century, and the most significant is learning how to steer a course towards a future that best meets the collective needs and aspirations of humanity. However, the process of building that future started eons ago. It is reflected in the genetic and ideational heritage of the human race, and in the life of the institutions that we have created. We are at a new stage of evolution—one that has transitioned from individual and group selection to institutional, global and potentially galactic. The fitness landscape is no longer determined by the natural world but by the human one. In order to survive and thrive we need to identify and promote institutional behaviors that satisfy our human needs and aspirations.

It is imperative that we continue the enterprise of scientific inquiry. Human civilization should remain committed to the pursuit of knowledge about our world and how to continue making it a better place for us and the generations that follow. This will power the spear of human civilization. However, we also have to foster institutions, the networks of human civilization, including science itself, that work effectively together and that embody human empathic qualities. We must design the fitness landscape for human institutions to reinforce the qualities of trust, honesty, mutual respect, humility and shared commitment. In short, we should arm the tip of the spear with love in its most universal form.

References

1. de Chardin, P.T.: The Evolution of Chastity (1934). as translated in Hague, R. (ed.) Toward the Future. http://en.wikiquote.org/wiki/Pierre_Teilhard_de_Chardin (1975) (last downloaded 4-6-2014)
2. Wrangham, R.: Catching Fire: How Cooking Made Us Human. Basic Books, New York (2009)
3. See, for example, the research of Angus Maddison on increases in historical GDP per capita reported on http://www.theworldeconomy.org (last downloaded 4-6-2014)
4. Rees, M.: Our Final Hour, A Scientist's Warning, Paperback edn. Basic Books, New York (2003)
5. The Era of Darwinian Evolution is Over. Digital New Perspective Quarterly. http://www.digitalnpq.org/articles/nobel/193/07-23-2007/freeman_dyson (2007) (last downloaded 4-6-2014)
6. Jeffrey, G.: Emergence as a construct: history and issues. Emerg.: Complex. Organ. 1(1), 49–72 (1999). http://www.complexityandsociety.com/files/4413/1692/0252/Emergence_as_a_Construct--History_and_Issues.pdf (last downloaded 4-6-2014)
7. Mitchell, M.: Complexity: A Guided Tour. Oxford University Press, New York (2009)
8. Smolin, L.: The Trouble With Physics. Mariner Books/Houghton Mifflin Company, New York (2006)
9. Gleiser, M.: A Tear at the Edge of Creation. Free Press/Simon and Schuster, New York (2010)
10. Wilson, D.S.: "Clash of Paradigms" in the Social Evolution Forum, 13 July 2012. http://socialevolutionforum.com/2012/07/13/david-sloan-wilson-clash-of-paradigms-why-proponents-of-multilevel-selection-theory-and-inclusive-fitness-theory-sometimes-but-not-always-misunderstand-each-other/ (last downloaded 4-6-2014)

11. See generally, The Journal of Economic Issues (ISSN: 0021-3624), published quarterly by the Association For Evolutionary Economics, Salisbury, NC
12. Dawkins, R.: The Selfish Gene. Oxford University Press, Oxford (1976)
13. Nowak, M.: Supercooperators. Free Press/Simon and Schuster, New York (2011)
14. Haidt, J.: The Righteous Mind. Vintage Books/Random House, New York (2012)
15. Wilson, D.S.: Darwin's Cathedral, University of Chicago Press, Chicago (2002) and Wright, R.: The Evolution of God, Little, Brown and Company, New York (2009)
16. Support for this conclusion is found in several of the works previously cited. For example, Mitchell, op cit., discusses the scaling structures that appear in a variety of complex systems
17. Taleb, N.: The Black Swan: The Impact of the Highly Improbable, 2nd edn, p. XXVI. Random House, New York (2010)
18. SETI Institue, FAQs, http://www.seti.org/faq#csc2, last downloaded 4-6-14
19. http://www.nasa.gov/mission_pages/msl/news/msl20121203.html, last downloaded 4-6-2014
20. http://en.wikipedia.org/wiki/List_of_interstellar_and_circumstellar_molecules, last downloaded 4-6-2014
21. Livio, M.: On Extraterrestrial Life, 12-5-13. http://www.huffingtonpost.com/mario-livio/on-extraterrestrial-life-_b_4391219.html, last downloaded 4-6-2014
22. Rose, D.: The Moral Foundation of Economic Behavior. Oxford University Press, New York (2011)
23. Keim, B.: The Secret Life of Everything: Where Your Stuff Comes From, 29 October 2013. http://nautil.us/blog/the-secret-life-of-everything-where-your-stuff-comes-from, last downloaded 4-6-2014
24. Bacon, F.: From The Advancement of Learning, paragraph I(3), originally published in 1605; posted on Project Gutenberg at http://www.gutenberg.org/dirs/etext04/adlr10h.htm, last downloaded 4-7-2014
25. The Ten Commandments, from the Book of Exodus, Chapter 20. (ref: http://www.biblegateway.com/passage/?search=Exodus+20, last downloaded 4-7-2014)
26. The Book of Mark, Chapter 12, verse 31. (re: http://www.biblegateway.com/passage/?search=Mark%2012:29-31, last downloaded 4-7-2014)

Biography of Author

George Gantz is a retired business executive with a life-long passion for mathematics, science, philosophy and theology. He has a Bachelor of Science degree with Honors Humanities from Stanford University and now directs an internet Forum on Integrating Science and Spirituality (http://swedenborgcenterconcord.org/) and blogs on related topics.

Chapter 14
U-Turn or U Die

Flavio Mercati

> *In their explorations, they encountered life in many forms and watched the workings of evolution on a thousand worlds. They saw how often the first faint sparks of intelligence flickered and died in the cosmic night.*
> *And because, in all the Galaxy, they had found nothing more precious than Mind, they encouraged its dawning everywhere.*
> Arthur C. Clarke, 2010: Odyssey Two.

14.1 Who's the Stonehead Now?

A lot has been written about the tragic story of *Rapa Nui* (Easter Island). A thriving culture, capable of building hundreds of *moai*, the emblematic giant stone statues, which obliterated itself due to unsustainable practices, mainly deforestation and overpopulation. The people who cut the last palm tree on the island are often mentioned. Did they understand they were committing suicide? Or were they just too concentrated on other things which they misjudged as more important than trees?

The giant stoneworks which haunt the island don't look promising in that respect. One feels a strong temptation to associate them with the ecological disaster that destroyed Rapa Nui's civilization. Were those people so obsessed with building titanic stone heads that they didn't hesitate to murder their environment—and their society with it—to build more?

Several pieces of evidence frighteningly point in that direction. Virtually all of the standing statues have been toppled. Moreover only a quarter of the statues were installed, nearly half of them remained in the quarry. It looks like there was a fierce

F. Mercati (✉)
Perimeter Institute for Theoretical Physics, 31 Caroline Street North,
Waterloo, ON N2L 2Y5, Canada
e-mail: fmercati@perimeterinstitute.ca

© Springer International Publishing Switzerland 2016
A. Aguirre et al. (eds.), *How Should Humanity Steer the Future?*,
The Frontiers Collection, DOI 10.1007/978-3-319-20717-9_14

145

civil war in which people unleashed their fury against the statues, as if they were rebelling against the gods or authorities who let them down. And it looks like this was a sudden disaster that struck an unaware population. The Rapanui were in the process of building *three times more* moai than there were in the whole island: the peak of an exponential growth. And there is strong evidence that this cataclysm happened just about when the last tree was cut. Rapa Nui's history has attracted much interest in the last decades. It became popular because it sounds like a warning to our current civilization. It is not my job to argue about the similarities between that insular society and the modern global one, as it has been done very effectively by more authoritative voices [1]. The metaphor is clear: The Rapanui are us. The palm trees are our environment. The island is planet Earth.

But are we as foolish as the Rapanui? My present impression is that yes, we are at least as foolish and as capable of harming our island-planet. In fact, we have our giant stone heads, which we stupidly value more than the environment. You name it: nuclear submarines, financial services industry, fighter aircrafts, show business, intercontinental ballistic missiles... the list could continue forever. These are things which we value immensely—one could argue, given their monetary values, that they are the things we value the most, as a society. And yet their value for our survival as a species varies from virtually useless to outright menace.

We are not doing this in a moment of abundance, with vast resources at our disposal. For the first time in its history Western Civilization is starting to see the toll that its development (and human activities in general) has been taking on the planet. And it might already be too late.

Jared Diamond [1] lists 12 environmental problems which, if not tackled, will become critical within 50 years. Each one of them, if not solved, has the power, alone, to destroy our civilization:

- Deforestation/habitat destruction
- Soil erosion, salinization, fertility loss
- Water management problems
- Overhunting
- Overfishing
- Effects of introduced species on native species
- Overpopulation
- Increased per-capita impact of people
- Anthropogenic climate change
- Buildup of toxins in the environment
- Energy shortages
- Full human use of the Earth's photosynthetic capacity.

It really looks like we're cutting our last tree, while obsessed with making war on each other, idolizing celebrities and people with obscene concentrations of wealth. Some of our grandest projects, like Dubai's themed housing mega-projects, risk lying unfinished in the construction site, gloomy monuments to futility in a world in ruins,

Fig. 14.1 A moai from Rapa
Nui (source: Flickr user
TravelingOtter)

like Rapa Nui's largest Moai shown in Fig. 14.1. I think it is fair to say that, with
all our knowledge and technology, we are more of a bunch of stoneheads than the
Rapanui.

14.2 Give Me a Reason

I have no original solutions to humanity's problems to offer in this essay, no more
than all the people who devoted their life to saving us from ourselves. As Diamond [1]
puts it:

> We don't need new technologies to solve our problems; while new technologies can make
> some contribution, for the most part we "just" need the political will to apply solutions
> already available.

My main goal in this essay is to present a different perspective on the issue. In particular, I will offer yet another *reason*, which is seldom thought of, why we should solve our environmental problems.

The passage from Arthur C. Clarke's *2010* I quoted at the beginning depicts an ancient and wise spacefaring civilization which, having gained an almost total control over matter and energy, considers life and in particular intelligent life the most precious thing in the Universe.

I'm not as enlightened as Clarke's ancient civilization, so the—probably real— superior value of intelligent life for our species is still not clear to me.[1] But I do clearly see the inestimable importance of life for our very survival. And I'm not talking about *any* life. I'm exclusively talking about the only kind of life that, by all chances, is useful to us: Terrestrial Life, namely life on Planet Earth.[2]

The main point I want to make with my essay is:

WHAT WE CONSIDER VALUABLE NOW, WE WILL LAUGH AT IN THE FUTURE.

this, of course, provided we survive. And the main conclusion I want to draw is:

WE NEED A U- TURN IN OUR COURSE, AND TO INVERT OUR SCALE OF VALUES.

The things we crave the most now are those things we believe bring us wealth. They are above all primary resources: oil and other fuels, radioactive elements, rare earths, ores, water. They are among the simplest compounds that matter can form. There is a precise reason for that: our backwardness. Our simple technology only allows us to extract energy or forge into building blocks very simple arrangements of matter. We've gotten progressively better at that, and we're still improving, but we're still hugely dependent on these simpler and more versatile states of matter. We have little use for more complicated naturally occurring compounds, like most rocks. And we have an almost complete disregard for the most complicated of all compounds: living matter. We have learned how to control a tiny fraction of it (domesticated plants and animals, and some bacteria and fungi, which represent a negligible portion of living species), but it appears we have little or no use for all the wild life. This is illustrated by the fact that we have replaced most of the biomass of land mammals on earth with domesticated ones (see Fig. 14.2).

Incidentally, these compounds we are obsessed with are extremely abundant in our Universe. Thanks to their simplicity, they are by far the most common things you can find in space: hydrocarbons are ubiquitous in the atmospheres of gas giants, rocky planets are teeming with heavy and radioactive elements.

---

[1]Maybe one day we will reach that stage at which we will understand the importance of intelligent life as well, and we will cherish, say, dolphins much more than caterpillars, *and not because they're more cute* (A/N: any reference to D. Adam's *The Hitchhiker's Guide to the Galaxy* is purely coincidental).

[2]Extraterrestrial life, or life that evolved *independently* on other planets, won't share our basic biochemistry and will almost surely be either completely inert or utterly poisonous to us: imagine lifeforms which are based on arsenic, or the lifeforms that evolved on the surface of a neutron star described by R.L. Forward in *Dragon's Egg*. We won't be able to integrate with an extraterrestrial living environment, and we better stay away from it.

Fig. 14.2 Pie graph showing the total mass of different land mammals. Wild mammals (represented by an *elephant*) represent only a tiny fraction of the overall mass, dominated by domestic animals and humans (source: explainxkcd and [2]). According to Smil [2] at the beginning of the 20th century the biomass of wild animals was equal to that of humans, while now it is less than one tenth

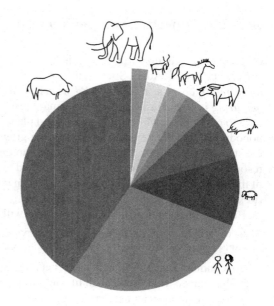

We consider them precious because we're stuck on Earth, and we only have extremely primitive means of extracting them.[3] If we were smarter, we would have access to the cornucopia that is waiting up in the sky, and we would never think about poisoning our own land to extract any of that stuff. On the other hand, the living creatures we treat with such contempt here on Earth are very rare in the Universe. We have still no evidence of extraterrestrial Life, and even if most scientists would bet that it abounds in our Galaxy, we can be sure of one thing: it's by no means as abundant as the simpler inanimate compounds we're so addicted to. And the situation is, in my opinion, much more severe than that: if we restrict to *Terrestrial Life* (life as we know it on Earth), all the chances are that it's basically unique. You can forget about the little green men because, if life evolved truly independently on each planet, there won't be anything like that out there in the Universe. Extraterrestrial life will be wildly different from what we're used to. I'm sure that, in its beauty and strangeness, it will beat by a long shot our craziest imaginations. This is what Nature always does when she lets us peek through her mysteries. And with all that strangeness, incompatibility will come. I'm convinced that extraterrestrial life will be so radically different from what we're used to that, for the purposes I'm considering her it will also be utterly useless to us. So here comes the consideration that motivated this essay:

TERRESTRIAL LIFE IS EFFECTIVELY UNIQUE IN THE UNIVERSE.

So here's my thought: aren't we squandering, simply through ignorance of its objective value, a very precious resource in which we were initially immensely rich?

---

[3]In most cases we just dig giant holes in the ground, grind the rocks we find, and filter the desired material away, devastating the environment in the process and leaving an incurable wasteland behind.

14.3 The Objective Value of Terrestrial Life

Of course there is no surprise in this 'inversion of values' that holds sway in our current society: as Adam Smith put it [3]: *The real price of every thing, what every thing really costs to the man who wants to acquire it, is the toil and trouble of acquiring it.* In the last century, the primary resources we depend on have been very hard to get hold of, while wild animals and plants were abundant and pervasive, and nobody saw much use in them. Should we grow out of our infancy and reach for the stars (or at least the Solar System), the immense disparity in abundance of those resources compared to Terrestrial Life is bound to shift the balance of value in favour of the latter. This just out of sheer scarcity, even if we remain blind to the "use value" (to keep using the economic language) of life.

But the situation is actually different, as I will argue now:

IF WE SURVIVE THE ENSUING THREATS TO CIVILIZATION AND WE BECOME A SPACE-FARING SPECIES, THE UTILITY OF TERRESTRIAL LIFE FOR US WILL BECOME IMMENSE.

To conquer space, we will need habitable planets to spread to. Potentially-habitable planets won't look at all like immense prairies teeming with native populations to mass-murder and bisons to kill and replace with cattle. They will look more like Fig. 14.3:

Mars is a terrible place to live, at the moment. But if you exclude the Earth, it's probably among the best places for us in the Universe. It is this sort of planets we will be looking for when we'll be searching for new homes. We won't go after the planets with life: as I said, they will be full of life-forms with a totally incompatible biochemistry, likely poisonous to us. You don't want to eat a silicon-based fruit, or breathe sulphur dioxide. If we were to turn an already-inhabited planet into an inhabitable place for us, we would be forced to completely sterilize it, killing every living thing in the process, and bring it back to a state similar to that of present-day Mars. Our best choice, instead, is to go for planets in the 'Goldilocks zone'[4] with sufficient mass to retain a long-lived atmosphere, with little or no native atmosphere and with a readily-accessible source of water in great amounts.

On this sort of *tabula rasa*, the humans of the future will first have to install an atmosphere and oceans, and then build an ecosystem from scratch. If the first task looks titanic to the contemporary eye, the second one is gargantuan. The challenges posed by a fundamentally alien environment to Earthly organisms are such, and the variables involved are so many, that this endeavour requires a much more mature science than ours. For such a feat, a vast choice of tools is absolutely necessary.

Fortunately, we have at our disposal in this very moment on Earth the most amazing collection of such tools, featuring more than 5–10 million (some say 100 [4]) different instruments. This super Swiss army knife is our ecosystem, and the tools are the species of living organisms that compose it. Given a virgin planet, we could play endlessly putting together all sorts of Terrestrial species to try and form a functioning

---

[4]The region around a star within which planets with sufficient atmospheric pressure can support liquid water at their surfaces.

Fig. 14.3 *Curiosity*'s picture of twilight on Mars (Image courtesy: NASA)

ecosystem that adapts well to the conditions on that planet. If, in a far future, we manage to master this complex art, no one can predict now what sorts of organisms living today will turn out to be useful—or even essential—for the survival of those engineered worlds. Earth will be an invaluable repository of genetic diversity, and virtually all the species presently living here are sure to find an application, sooner or later, somewhere in the Galaxy.

So we need to save the biodiversity of Earth. But, tragically, we've been doing exactly the opposite for a long time. We've been blindly wiping out our ecosystem since the recent advances in our hunting techniques, which took place 50,000 years ago. And we've been doing it at an ever-accelerated pace: We've already extinguished virtually all of the megafauna in 4 continents out of 5, and one half of Earth's higher forms are expected to get extinct within 100 years [5]. According to May [4]:

> recent extinction rates in well-documented groups have run 100–1000 times faster than the average background rates. This is the same acceleration in extinction rates as characterizes the Big Five episodes of mass extinction in the fossil record. And four different approaches to estimating impending rates of extinction suggest further acceleration by a factor 10 or more.

Maintaining the biodiversity of planet Earth is one of the necessary conditions for our success in the colonization of space. We've been wasting this biodiversity to such an extent that we are at risk of losing it completely in a few decades. All evidence at this stage suggests that we've already triggered the Sixth Mass Extinction in the history of Planet Earth.

14.4 Что делать (What Is to Be Done?)

After such a pessimistic note, I want to conclude this essay with a bit of hope. If I were convinced that everything is lost, there would be no point in writing this essay.

I explained in the last two sections why we should preserve biodiversity, as it will be our most precious legacy to future generations trying to colonize other planets. I will now stress some things we should do, in my modest opinion, to avoid the ecological disaster we are facing. I will conclude with a picture of the kind of society we should be working to create, one in which biodiversity is kept as the most cherished treasure for the use of future generations.

So I think there is still hope, but, as my title suggests, we need a sudden and dramatic change of course. It's absolutely not enough to gently adjust our trajectory, as most lawmakers seem to believe. It's not even enough to hit the brake: even if our development came to a full stop now, that would probably not arrest the ongoing ecological collapse. We need to make a U-turn, and start operating actively to heal our environment.

Most ecosystems are already compromised, some of them irreversibly. On most of them (think about the oceans) we have little or no data. They are collapsing, and in most cases removing the original causes of their deterioration won't prevent, at this point, their demise. We need to learn how to *rebuild* them, understand their workings and fix the weak links of their ecological networks. This task requires an immense amount of work, both theoretical and applied. Considering what is at stake, it is just staggering that we are not already investing most of our resources on this. By contrast, the vast majority of our resources are devoted to building our modern versions of moais. Our task is not only difficult: it's also dangerous, because we need to make a U-turn at full speed, and there is a serious risk of losing control of the vehicle. Tampering with the ecosystem without the right knowledge might do more damage than good. But we have no other choice: we have one chance, and we better get it right.

14.4.1 Reduce Our Impact

Population growth is often cited among our worst problems. While it is certainly true that we're still growing and this is a factor in the increase in our impact on the planet, at this point it is not the most important factor anymore. In fact the last decade saw *"a 'tipping point' in our demographic history: the planet's average woman is having almost exactly one female child [...] if indeed fertility rates continue at replacement levels the population will continue to grow, to some nine billion or so, before possibly coming to equilibrium around 2050"* (from [4]).

The population growth problem is being tackled, and that's good news. But the burden we are putting on the planet's ecological resources is still growing at an alarming rate. In Fig. 14.4 (left) I show an estimate of WWF's indicator called 'Ecological Footprint' of humanity, expressed in terms of the number of Planet Earths that would be needed to stably sustain human consumptions. It was already well above 1 Planet Earth in 2002.

What we really need to do is to reduce the *per-capita* burden on the environment, starting with already-developed countries. In Fig. 14.4 (right), again from WWF, I show the Ecological Footprint of different macroregions, compared with their capacity. It is clear who is living above their possibilities. The main emergency, at the moment, is the increase in consumption of developing economies. These people aspire to reach the living standards of the European Union or North America, with

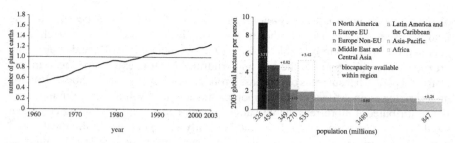

Fig. 14.4 (*left*) Estimated "Ecological Footprint" of the human population between 1960 and 2003. This indicator measures the global hectares of land that are needed to sustain the human population at its actual consumption levels. This graph is expressed in units of the total capacity of planet Earth [6]. (*right*) Ecological Footprint in 2003 of each of the planet's major geographical regions, expressed in global hectares per person. The *horizontal axis* represents the populations of those regions, and the *rectangular areas* gives the total Ecological Footprint per person. The *dashed rectangles* represent the maximum capacity of each region [6]

corresponding levels of consumption and waste. They are getting there at an alarming speed. As Diamond [1] wrote in 2005: "*Even if the people of China alone achieved a First World living standard while everyone else's living standard remained constant, that would double our human impact on the world*", and that figure goes up to *12 Earths* if all of the population of what used to be called "The Third World" adopted "First World" living standards.

Nobody in the west is entitled to blame China or Brazil for aspiring to an unsustainable lifestyle, as long as North America and the EU keep exploiting, respectively, 3.7 and 2.6 hectares per person more than the maximum they would be entitled by their land (cfr. Fig. 14.4). It is therefore in the western world we need to intervene more urgently, starting with North America, by far the biggest waster. If we find a viable compromise between wealth and sustainability in this region, this would set forth an example for the kind of development that other regions ought to pursue. Again, it is not my job here to offer a solution, other people have done a much better job over many years than I could possibly do here.[5] Moreover, there is no one-stroke solution: the problems are complex, and the number of different interventions needed is large.

It is true that a lot has been done, in western countries, since the 70s: for example the emissions of internal combustion vehicles have been hugely cut, the oil industry has reduced its impact,[6] the management of waste and garbage has improved a lot, and the overall ecological awareness of people has increased from zero to something. However, people are still not properly informed about what needs to be done, there is only a generalized awareness of impending disaster. The U-turn I'm advocating must be first of all in mentality. For example, North America is still a huge consumer

---

[5]Just take a look at WWF's website, John Baez's Azimuth Project, or the Worldwatch Institute to mention some.

[6]Unfortunately we cannot say the same about all kinds of extraction industries.

of meat, which is an extremely inefficient source of sustainment because all the land we use to raise cows and other livestock could be used to support much more people if devoted to purely arable farming (without mentioning all the damages that grazing livestock imparts to the soil and wildlife). Meat is extremely cheap in this historical period because it is being mass-produced with industrial methods.[7] But this underpricing is an artifact of the temporary 'inversion of values' we're living with. If we attributed a more reasonable value to land, ecosystems and resources, meat would be a luxury good. And that would actually be a good thing: the meat-based diet of Northern Americans is very unhealthy, supporting the largest concentration of obese people in the world.

Here's a radical proposal, which might turn out impossible to realize, on how to reduce our Ecological Footprint, improve our health and quality of life, and the beauty of our lands: let's give up animal husbandry altogether. I'm not advocating an impossible mass-conversion to vegetarianism: we would still be consuming meat, but only game meat. It seems very unlikely that people would accept this. And yet our ability to adapt to radical changes will determine our chances to survive the incoming crisis.

14.4.2 Let's All Become Tikopians

In his book *Collapse*, Jared Diamond mentions some pre-industrial societies which, out of trial-and-error and a good deal of luck, managed to solve their ecological problems. One of the most striking examples is a tiny Pacific island called Tikopia, isolated by hundreds of miles of ocean from the nearest land. With less than 5^2 km of surface, Tikopia supported a population of 1,200 people (very dense for a pre-industrial society) continuously for almost 3,000 years. Here is Diamond's description of the island:

> As you approach Tikopia from the sea, the island appears to be covered with tall, multi-storied, original rainforest, like that mantling uninhabited Pacific islands. Only when you land and go among the trees do you realize that true rainforest is confined to a few patches on the steepest cliffs, and that the rest of the island is devoted to food production. [...] This whole multi-story orchard is unique in the Pacific in its structural mimicry of a rainforest, except that its plants are all edible whereas most rainforest trees are inedible. [...] Sustainable exploitation of seafood resulted from taboos administered by chiefs, [...] (which) had the effect of preventing overfishing.

Tikopians traditionally maintained their population stable by means of contraception, abortion, infanticide and ritual suicide (or emigration by sea, which was virtually equivalent to suicide). The arrival of Europeans introduced religious or cultural

---

[7]Many people are distressed by the industrial methods of producing meat: meat factories and industrial butcheries look a lot like the animal version of Nazi extermination camps. I might be sympathetic with these feelings, but I think we have a much better basis than the moral one to argue against these practices.

taboos against most of those practices and triggered an overpopulation crisis and consequent famine in the 1950s. After that,

> ...Tikopia's chiefs limit the number of Tikopians who are permitted to reside on their island to 1,115 people, close to the population size that was traditionally maintained by infanticide, suicide, and other now-unacceptable means.

The Tikopians successfully went through the U-turn and inversion of values I'm advocating. They started practicing slash-and-burn agriculture, overexploitation of bird and marine life and damaging pig farming. Gradually, over thousands of years, they learnt to carefully safeguard wildlife, they decided to slaughter all pigs and they stopped destroying the forest, make it instead into their primary resource. We need to find a way to do the same on a much larger scale: our values need to be aligned with the factual importance that the different resources have for our survival. Fostering the regrowth of our biosphere must become economically advantageous, the wealth of the people must turn into a synonym of the health of their environment. To do so, we don't have millennia like the Tikopians. We need to be swift. You decide. You turn, or you die.

Acknowledgments I am very grateful to Julian Barbour for his help improving the language of this essay. I thank also Matteo Lostaglio for his useful observations (and typo-spotting), which helped improving the text. Finally I thank Niccolò Loret for alerting me to the unintentional reference to "The Hitchhiker's Guide to the Galaxy", when talking about dolphins and caterpillars.

Technical Endnotes

On the Feasibility of Surviving Without Nature

Reading my Sect. 14.3, one could be tempted to object: a future version of humanity so advanced that it moulds entire planets according to its needs surely won't depend on animals and plants for its sustenance! This, I think, is a misconception, fueled by decades of science-fiction which has accustomed us to expect marvels like synthetic food, self-aware robots, living creatures created genetically from scratch. These things are not impossible, but at the moment they definitely belong to the realm of the imagination. For example, in 70 years of research on Artificial Intelligence, we have made little progress since Turing's groundbreaking work. Similarly, our understanding of the workings of life is extremely rudimentary: even the simplest cells are an absolute conundrum, and we are just starting to make the first timid steps towards engineering a cellular membrane. It can probably be done, it will just take a very long time to understand how. Terraforming a planet, in contrast, requires less of a conceptual breakthrough, and could well be within the possibilities of the humans of tomorrow. Creating a biosphere will then be the next step. Every species, including ours, is defined by the complex and fragile interrelations that it maintains with its environment: from the viruses, bacteria and protozoa to the flora and fauna that surround it. Its survival depends on its equilibrium with this environment. It's not only the chemicals that are provided by exchanges with the environment: think about the fact that bacteria in our bodies outnumber human cells 10–1, and we depend on them in ways that we haven't completely understood yet. Our success in colonizing most regions of the Earth has given us a false sense of omnipotence, as if we could free ourselves from the bonds that tie us to our Terrestrial environment and reach for the stars, providing for all our needs on our own. This is plain wrong: our diffusion over the five continents proves nothing about our ability to adapt to them. We colonized most of these lands without finding sustainable practices to provide for our needs, and if we keep our present course we will rapidly get extinct in most of these regions, which cannot really support us. We're like bacteria in a petri dish that experience an exponential growth and assume they will do so forever. We badly need to learn how to survive—stably—on Earth, and only then we might look up towards our natural destiny: the stars.

On Abandoning Animal Farming

We could forbid the breeding of domestic livestock (or impose exorbitant taxes on it) and get all the meat from hunting wild animals in the forests. If we do a good job in recovering wild ecosystems, the forests will be able to sustain a reasonable, although greatly reduced, production of meat. Hunting will be necessary in any case to correct the unbalances and keep the ecosystems stable. The meat produced in this way would be way less, and consequently way more expensive. But hey, it's a luxury you have to pay for, and it's even bad for you. We consider acceptable to impose crazy taxes on cigarettes. I don't see the difference with heart-disease-causing meat. The current average per capita meat consumption in the US is a disturbing 377 g/day of beef, pork and poultry (2009 data), of which 191 g come from beef and pork alone [7]. I must remark that there is a lot of pressure, nowadays, on nutritionists trying to curb the consumption of red meat in the west. To me there is absolutely no controversy about the superiority of a low-meat diet, as is proven by the better health performance of countries that rely mainly on fish proteins like Japan, or champions of the Mediterranean diet like Italy.

Of course hunting should be strictly regulated (as it already is to a large extent, in western countries); we cannot allow our forests to be subject to the same ruthless overexploitation that is devastating the oceans. Actually we should apply worldwide severe regulations on fishing as well. The European Union is moving in the right direction by imposing fishing quotas, but these measures remain insufficient, with quotas set well above the levels recommended by scientists (source: The Guardian).

References

1. Diamond, J.: Collapse: How Societies Choose to Fail or Succeed, Revised edn. Penguin, New York (2005)
2. Smil, V.: The Earth's Biosphere: Evolution, Dynamics, and Change. MIT Press, Cambridge (2003)
3. Smith, A., Garnier, M.: An Inquiry into the Nature and Causes of the Wealth of Nations. Nelson, Edinburgh (1845)
4. May, R.M.: Ecological science and tomorrow's world. Philos. Trans. R. Soc. B: Biol. Sci. **365**(1537), 41–47 (2010)
5. Wilson, E.O.: The future of Life. Random House LLC (2002)
6. W. W. F. for Nature: WWF 2008 living planet report. www.panda.org
7. Food and A. O. of the United Nations: Food Supply. FAOSTAT. http://faostat3.fao.org/faostat-gateway/go/to/home/E

Chapter 15
Smooth Seas Do Not Make Good Sailors

Georgina Parry

Abstract Smooth seas do not make good sailors. Rough and unpredictable ones, difficulties, problems, challenges do. Another premises on which the essay is founded is that the future can be built rather than just steering towards a space-time future that has existence in the space-time continuum, or toward a particular branching of the universe. The story further explores human ability to adapt and implement existing solutions to a large number of problems, and how our understanding of both physics and biology can be taken and utilized to preserve and improve our quality of life, health and dignity under extreme conditions. The story takes a look at a typical day for an inhabitant of one sanctuary and looks back at the attitudes of the past and also to the future. Having achieved a truly sustainable, self sufficient, symbiotic lifestyle migration to other worlds can now be contemplated. There is a narrative in black type which is describing life in the sanctuary. There is the writing on the girl's computer and home screen in **bold** type, sayings that bind the community together and lessons that the girl is being taught. Then in *italic* type is technical information from the Knowledge Hub, which I imagine is from the communities super computer and repository of knowledge. The technical information is not a necessary part of the story but provides actual scientific information to back up the story details. The tale progresses from a rather stark opening quote that shows the devaluation of humanity and ends on an uplifting quote that in contrast shows the unappreciated value of life, and especially the human being. It also progresses through a day from 'sunrise' to 'sunset'. Physics is woven into the tale both in the context of problems we will face and as solutions to problems. Nature has been taken into the sanctuaries for its continuation and for the needs of mankind. The people have a clear shared purpose, to preserve and propagate the tree of life. Due to certain aspects of human nature and personality some do not want sanctuary or can not be given sanctuary for the safety of the rest. The sanctuaries are however a way of allowing a large number of people survive rather than the human race going through an evolutionary bottle neck. While it remains optimist that a technological civilization can survive, it may also serve as a grim reminder of the difficulties we may face in our own uncertain future.

G. Parry (✉)
Nelson, New Zealand
e-mail: g.woodwardphysics@gmail.com

© Springer International Publishing Switzerland 2016
A. Aguirre et al. (eds.), *How Should Humanity Steer the Future?*,
The Frontiers Collection, DOI 10.1007/978-3-319-20717-9_15

159

Asimov: In the same way democracy can not survive overpopulation human dignity can not survive overpopulation, convenience and decency can not survive overpopulation. As you put more people into the world the value of life not only declines but it disappears. It doesn't matter if someone dies, the more people there are the less one matters [1].

There is no need for alarm clocks, as all are woken by the gentle artificial sunrise, as the lighting is gradually restored to the habitat areas. Optional birdsong, relayed live from the Wildlife Zones, plays during the breakfast hours.

Grace makes her own late breakfast with the ingredients her mother has left by the blender. Spiralina, chlorella, coconut milk, coconut oil, avocado. One fat metabolism activating, micro-nutrient packed, tasty, antioxidant rich, green, smoothie! She takes it to her private room. Then logs herself into school. Her friends avatars are already at their desks. The screen switches to the 'whiteboard'.

15.1 The First Law of Sustainability

15.1.1 Population Growth and or Growth in the Consumption of Resources Cannot Be Sustained

She wonders, fleetingly, why it is there every morning. It is said that the world's population within the sanctuaries is stable at 10 billion with 0 % growth. Disease, famine and war were left behind. Those with the calling to dedicate themselves to the life of warriors patrol the surface, keeping away criminals, terrorists and debris that pose a threat to the ventilation shafts and logistics portals. They collect weather data from the surface weather stations and carry out repairs on the apparatus when needed. They also escort culture and technology exchange missions between sanctuaries.

There are so many ways to serve mankind other than procreation. Those with the calling to be parents must apply for suitability examination and training. They give up a lot of benefits and freedom when making that choice of dedication. Besides, everyone can share in the lives of the sanctuary's children with regular updates on the 'C.F. (Child and Family) zone' channel. For some, watching their lives is like watching their own family. But most people are fulfilled, and feel significant and valued without children.

Grace remembers seeing the Affirmation on the entrance to C.F.

Every child a wanted child
Every child a healthy child
Every child loved and nurtured
Every child educated and socialized
Every child a blessing to our sanctuary

She had memorized it because it made her feel warm, appreciated.

New screen:

The greatest shortcoming of the human race is our inability to understand the exponential function.
Dr. Albert A Bartlett

Yes that one is familiar, the lesson of the bacteria in a jar.

From the Knowledge Hub

"Bacteria grow by division so that 1 bacterium becomes 2, the 2 divide to give 4, the 4 divide to give 8, etc. Consider a hypothetical strain of bacteria for which this division time is 1 minute. ....One bacterium is put in a bottle at 11:00 a.m. and it is observed that the bottle is full of bacteria at 12:00 noon". Dr. Albert A Bartlett
*When was the bottle **half-full**? Answer: **11:59 a.m.**!*
*How long can the bacterial growth continue if the total space resources are quadrupled? [I.e. three new bottles are found.] Answer: "Quadrupling the resource extends the life of the resource by only two doubling times! [**Just two minutes**]. When consumption grows exponentially, **enormous** increases in resources are consumed in a very short time!"* Dr. Albert A Bartlett [2].

These are important lessons that mankind had to learn and every child in the global sanctuaries education programme learns.

That example brings to mind the symbiont microorganisms used to defend the sanctuaries from disease. Antibiotics are mostly obsolete, as are disinfection and sterilization. Disease is avoided by maintaining healthy bodies, inside and out, and by having a healthy living symbiont environment. Cleansing is carried out with mild cleansers containing encapsulated beneficial microbes, such as Lactobacilus species. These symbionts compete with pathogens, preventing their proliferation.

New Screen:

There is no wealth but life. He who buys what he does not need steals from himself.

She has seen this affirmation before, at the start of the history class last term. She recalls being told that growth of consumption and production used to be a measure of success. How silly that was. Now even infants begin to learn about the exponential function, to understand that limits can not be exceed. At first they think being allowed to take more and more jelly beans out of the box is a good thing but when the box is empty they gradually begin to understand.

It's like this everyday. It's called social affirmation. Everyone in the sanctuaries shares this belief system, knowledge of the mistakes of the past, and of the sacrifices of the future problem solvers who made the survival of so many possible. Some had been leaders, some scientists, some philanthropists and some just visionaries who could see what needed doing but needed others to make it happen. What they had in common was they had opened their eyes and heeded the change in the weather. Each region of the Earth had its own challenges and there was not one solution to suit all. The future problem solvers pooled their knowledge, expertise, research and development, and ideas. Their Motto was "Adapt to survive".

New screen:

The human species is the ultimate generalist, we learn, we adapt, we survive.

Yes we do, she thought. When she had been younger she had worried about the unwritten future and what would happen to the thousands of sanctuaries scattered across the globe. Her mother listened to her concerns and then said "*Smooth seas do not make good sailors.*" Grace now knew it as one of the affirmations. Her mother patiently explained that it meant. "Whatever happens we will learn from it and be better able to deal with similar problems in the future. We are becoming good sailors, Grace." She smiled. There is no shame in being wrong so long as we learn. People can even choose to alter their dedication if they have a new calling, though most are permanently sterilized by the age of 25.

Grace is looking forward to Bioengineering today. They have had several weeks to work on the behavior programming and testing of their virtual 3D models. The designs are being presented for the first time. She is hopeful that her design will do well in the upcoming trials. The prize is to have the design 3D printed and presented to the people of habitat 2739 in the community Knowledge Hub. From there the design specification will be shared with the sanctuary's Central Knowledge Hub and could be accessed by other sanctuaries looking for a solution.

It certainly meets the specification. It must solve a problem, and it must be a survivor. Useless designs or ones that break easily will be eliminated early. She has designed a planting aid that could work in the gardens or wildlife zones. It is based upon a hunting spider morphology. Its spider like legs dig a hole in the ground, then it will lay a seed, like an egg, before carefully covering it with the good earth. It will work all day or night with its lightweight energy efficient design. It is tough too, with chitin and keratin components and soft, self healing polymer hydraulic arteries.

This bioengineering project has been a labor of love. She has nurtured and modified its virtual form into its final solution. It is a survivor, she is sure of it. Her confidence comes from long experience of design and programming life like forms and behaviors. As a toddler she had learned to use BAMZOOKi [3]. An old, old game but considered of educational value for pre-schoolers. Also, like all kids, she learned to read and learned to code at the same time. Collaboration on virtual habitat projects is a recess activity for the older children, when Kids Zone and Greenspace trips are not scheduled. The younger ones are nearly always given free play exercise in Greenspace. The older children's virtual selves, their avatars, can explore and play in the virtual worlds of their own creation, which may contain as much or as little magic as they wish to include. Sometimes it's pure escapism, sometimes testing realism. Either way it exercises their minds in play.

Passive Tunnel Ventilation Solution of Prairie Dogs

Application: Reducing the need for active ventilation or mechanical air conditioning of a subterranean network.

Bernoulli's principle can be regarded as a statement of the conservation of energy principle for flowing fluids. The qualitative behavior that is usually labeled with the term "Bernoulli effect" is the lowering of fluid pressure in

regions where the flow velocity is increased. Bernoulli's principle applies to various types of fluid flow [4].

Action of Bernoulli Effect on the Rim Crater Burrow Design of the Black Tailed Prairie Dog

When air flows across a surface the velocity gradient gives a potential source of work....The burrow of the black-tailed prairie-dog would be an extraordinarily large respiratory dead-space if diffusion alone were relied upon. Diffusion appears inadequate for sufficient gas exchange. The burrow is built for wind-induced ventilation. Typically it has two openings at opposite ends and mounds surrounding these, of two forms, one form on each entrance [5].

Entrances are of 3 kinds. Some have no mound. Others have a wide, round, unstructured mound (given the name dome crater), or a high mound, shaped like a volcano, with clearly visible rim, (given the name rim crater). If a tunnel has a low dome crater and a higher rim crater a breeze blowing across the ground will cause air to enter the lower dome crater and exit the higher rim crater [11]. Mounds with sharp rims are more effective exits for air than mounds with rounded tops. In black tailed prairie dog colonies the shape differences are correlated with the differences in height [6, 7].

The prairie dogs look cute, thinks Grace. Do they have prairie dogs in this sanctuary? She will ask the Knowledge Hub at the end of this lesson.

As many species as possible have been spared but many now exist only as records in the Knowledge Hubs, or in some cases as preserved genetic material. Plants and animals that have medicinal, companionship, bio-mimicry or bioengineering relationships to mankind have a special place of respect in the hearts and minds of the people. Each has been scrutinized for direct uses or adaptations that could be useful 'in the unwritten future'. Their images are displayed on the walls of the Knowledge Hub, together with the tree of life to which all Earth life belongs, as a mark of respect. The human population could not survive alone and these are fellow survivors, on the same journey, sharing the same key of life, the same genetic cypher.

New screen:

Desert snails-survivors [7].

****

New Screen:

Revision of Gills

Centrifugal Air Extraction Current application, auxiliary air supply for sub-aquatic sanctuaries, and submarines.

The separation of dissolved air from the water can be achieved by applying low pressure. Henry's Law states that, the amount of gas dissolve-able in a liquid is proportional to the pressure on the liquid. Reducing pressure causes dissolved gas to be released. Achieved by using a centrifuge causing lower pressure in

the center from where the air supply is collected. Developed by Like a fish Technologies [8].

Oxygen Diffusion Extraction

What are the human metabolic parameters that made nano-material diffusion extraction of oxygen problematic? Include metabolic O_2 requirements, toxicity of CO_2 and excess O_2, and the solutions to these problems in your discussion [9, 10].

Gills, grown according to nature's specification, with external placental interface are a great adaptation for the sub-aquatic sanctuaries. The sea floor was colonized when it was realized that sea rise could not be halted or held back. The sub-aquatic sanctuary dwellers are not confined now. They can swim unhindered in the ocean, and in the aquaculture ponds and aquatic wildlife habitats. It makes salvaging resources and history from the submerged cities a much more pleasant dedication. So cool, thinks Grace. She hopes one day to be accepted for a cultural exchange expedition.

The affirmation had become: Adapt to the environment. Do not cause more problems than you solve." (Bend like the willow.) Some people like to wear willow tree emblems a reminder of their strength without resistance to change, though others prefer the tree of life symbol, which is a reflection of their culture of bio-mimicry [11] and respect for life.

Back in history people had been saying: Adapt the environment to survive, but it was decided that geo-engineering projects were too dangerous because of the unknown, undesirable consequences that might occur [12]. Restoring the African grasslands and other areas of desertification has been a success though. Regenerating some habitat suitable for surface living and slowing CO_2 rise. Where once overgrazing had been blamed for desertification, a future problem solver, Allan Savory, realized it was insufficient beneficial trampling and manuring of the land that was the real culprit. Large roving herds of animals encourage plant growth with stabilizing root structure. When trampled into the ground, this plant material locks in the carbon. Savory had said "We must use livestock, bunched in very large moving herds, mimicking the way they used to roam when wild, or as they were herded in our agricultural past" [13].

It had taken time for the project to gain momentum. People had first to accept the paradoxical wisdom.

Recess
****

After shared recess, real or virtual play, the rest of the day is for personal research and study or creativity. Grace is building her own textures library. She likes to play the textures collection as a screen saver. Watching them slowly morph into each other as her mind relaxes and zones out. The project has approval from her parents and mentors. Stress reduction is seen as activity worthy of dedication time. It increases resistance to disease and improves mental and social well being. For the same reason some of the children choose to spend time with companion animals [14].

On the family's habitat screen is written -

Diversity aids survival, do not let the book be destroyed to save the page.

That's a historic affirmation. It had been necessary to prevent mass migrations, to protect civilizations and preserve unique and valuable cultures. At a time when the whole World was facing its own problems there had had to be global co-operation. To build sanctuaries to protect the lives of the helpless hordes, preserve the civilization that lay in the path of their migration, preserve the diversity of humanity and to live up to what it means to be human.

The 3D printing revolution [15] allowed rapid construction of temporary sanctuaries, which could then be extended and improved with a safe population. Some would choose the dedication to sanctuary design or dedication to construction. The production of new materials had also been a great help. Starting with the first self healing concretes [16] and self-healing polymeric materials [17, 18] but progressing to living materials, often chosen for their properties of self healing and strength. Bones with artificial supply of nutrients and oxygen. Living self healing echinoderm based skins, or chitin exoskeleton for biospheres. Organic forms of seed pods and plankton are preferred as dwellings [19] and work spaces. They make the artificial environment beautiful and seemingly natural. Humans are becoming symbiotic residents of living man made hosts. The life of the sanctuary depends upon its ecosystem of inhabitants and the inhabitants depend upon the sanctuary's survival.

Not everyone embraced the idea of living in the sanctuaries. Some stayed outside. They said they were waiting for God. Grace had asked her mother what that meant. She replied with the story of the drowning man. "Three ships came by but the drowning man sent each of them away. When the man died and met his God he asked, "why didn't you save me." His God replied, "I sent three ships." They had a certainty in their minds that couldn't be shaken. We're not prisoners, Grace, we choose this life. Those that want to leave often choose to serve the sanctuary with the dedication of warriors. But it is a very tough life. Most people wouldn't be able to survive out there. Without the sanctuaries programme mankind would have been facing an evolutionary bottle neck. That means nearly everyone dies and the few survivors become the progenitors of the new hominid species. Who knows if we would even recognize them as men, Grace?" [20].

****

The news-Weather patterns shifting again.

The weather has been temperamental for a long time. Long-term weather forecasting is difficult because it shows"sensitive dependence on initial conditions." Still called the butterfly effect here (Even though the reports from the sanctuary's warriors are that no butterflies have been seen on the surface for a long time!) It makes planning expeditions difficult.

From the Knowledge Hub

The Lorenz Strange Attractor was discovered while Edward Lorenz was trying to create a model of the atmospheric dynamics of planet Earth, in the 1960s He used a shortened version of the Navier-Stokes equations. The Navier–Stokes equations

describe the motion of fluid substances. These equations are found by applying New-
ton's second law to fluid motion, assuming that the stress in the fluid is the sum of a
pressure term and a diffusing viscous term (proportional to the gradient of velocity)
which describes viscous flow. The Navier–Stokes equations, in full and simplified
forms help with the design, study,modeling and analysis of many things involving
flow of liquids or gasses [21].

The non-linear dynamic system used for Lorenz' model illustrates cyclic long term
behavior revealing a hidden order [22].

Grace reasons that people must have once thought that, because small initial
alterations to a linear system lead to small changes, so as long as the climate inputs
were kept small, everything was fine. That's how a set of completely deterministic
equations behave. The inputs have not been small though (It may have been the
additional methane released by the melting ice sheets and permafrost.) [23, 24]. *And*
the equations are nonlinear. Nonlinear systems may demonstrate amazingly complex
'chaotic' behavior. The entire Earth climate and weather system is chaotic [25].

This sanctuary might one day be buried under ice. The scientists say it's to do
with shifting of the ocean currents [26]. Grace isn't afraid. It's a completely self
sufficient biosphere now containing many different ecosystems that are needed for
sustainability. The inhabitants can live under the ice, burrowing up into it if needs
be. Or the sanctuary could be put into hibernation. There has been a lot of research
into hibernation and aestivation of animal species [27, 28] as part of the Global,
future migrations research and development policy. Not only are the dormant states
well understood, they are used as therapeutic measures. Careful control of body
temperature can lower a person into a dormant hypothermic state, useful for traumatic
injuries and surgery [29]. Fever therapy and aestivation therapy are used for curing
infections and for cancer treatment [30]. Both of those kinds of disease are thankfully
very rare nowadays. It is thought that is because of the compulsory sleep regime
and optimal vitamin D levels [31] within the population, obtained through diet as
there is insufficient exposure to natural sunlight via the solar tubes. Sanctuary 2739
might even be chosen for the first migration into space. Nowadays it is unthinkable
that people would go out alone into space. As unthinkable as chopping off an arm,
throwing it away and expecting it to survive. It can't because it's part of a greater
whole, just like mankind belongs with the tree of life. The sanctuary's biosphere has
been self sufficient and sustainable for at least 10 years. It can survive and support
human life independent of any outside assistance. Sometimes Grace hopes 2739 will
be chosen, "we *are* ready now", she tells herself.

Everyone loves the artificial night. After a long tiring day at their dedication they
let their bodies bathe in the natural melatonin. It has built up in their bodies gradually
as the wavelengths of the lighting are centrally reduced from blue to red. It is called
sunset and is said to mimic the change in wavelengths of light on the surface with
which our metabolism evolved.

From the Knowledge Hub

Near sunrise and sunset, the natural light approaches nearly tangent to the Earth's surface. The light's path through the atmosphere is so long that much of the blue and green light is scattered out, making the clouds the sun illuminates appear red [32].

Mammalian eyes are not just a part of the sensory system that produces images Additionally the mammalian eye detects changes in light irradiance leading to non-image forming light responses. Synchronization of the circadian rhythm's clock is one non-image forming, irradiance dependent response [33].

Exposure to light at night has been correlated to several types of cancer, diabetes, heart disease, and obesity. Exposure to light suppresses the secretion of melatonin, and lower melatonin levels are considered the link with cancer. Researchers have also linked short sleep to increased risk for depression and cardiovascular disease. Researchers who shifted the timing of the circadian rhythms of their subjects found that their blood sugar levels increased, giving a pre-diabetic state, and decreased levels of leptin, the hormone that makes people feel full after eating [34].

Any light of can suppress the secretion of melatonin, but blue light is the worst culprit. Different wavelengths of light were compared for melatonin suppression and phase shifting of the salivary melatonin rhythm. The shorter wavelengths of 470, 497, and 525 nm showed the greatest melatonin suppression, 65–81 % [35].

Sleep is compulsory as a population health measure. The Culture and Entertainment Zone closes before 'sunset' and people return to their own habitats. EM devices are centrally shut down so there are no distractions. People sleep better than they ever did on the surface, and are healthier as a result.

Once upon a time it was thought that all life was within a space-time continuum, where past and future were the same and everything that would happen was already written into its fabric. Many clever men (and women) agreed, though it made some people uncertain and uncomfortable. The clever men would show the mathematics and win the argument. Then it was found that it was a trick of the light. The answer had been written on a web site called FQXi [36]. The image mankind call 'the present' has been written in the light but the material future has not been built. Now it is the mission of people like Grace, and the human species, to build a future. Success will be measured by the contentment, health, altruism, high culture, and creativity of its people. As a species, Homo sapiens sapiens are hackers of nature's solutions presented by the tree of life, that has evolved over millions of years. It could not be clearer if there was **How to survive** in big bold letters written on every surviving life form. Life provides lessons on, for example how chemistry and physics are harnessed efficiently. The solutions are applied to solve humanity's problems, or stored for future use. Many life forms, having 'sailed rough seas', have lessons to teach. Together the humans are becoming 'sailors' who can survive what ever nature's temper. Mankind is no longer *as* vulnerable, having developed a way of living, learned form the book of knowledge, written in the genetics, morphology, anatomy, physiology, metabolism, biochemistry, biophysics, behavior and ecology of life. A truly sustainable, versatile and adaptable way of life that is ready to be transplanted to other hostile worlds. The time and work it has taken to nurture and develop a human culture based upon sustainability, symbiosis, bio-mimicry, and respect for life has been preparation for the mission to propagate the tree of Earth life; so that even if all life on Earth is destroyed, G=C, C=G, A=T, T=A lives on.

Carl Sagan: A blade of grass is a commonplace on Earth; it would be a miracle on Mars. Our descendants on Mars will know the value of a patch of green. And if a blade of grass is priceless, what is the value of a human being? [37].

References

1. Asimov, I.: Interviewed by Bill Moyers on Bill Moyers' World Of Ideas, 17 October 1988, transcript p. 6. Reference from Wikiquote: http://en.wikiquote.org/wiki/IsaacAsimov, 14 April 2014
2. Bartlett, A.A.: www.albartlett.org/articles/art_forgotten_fundamentals_part_4.html. Accessed 14 Apr 2014
3. Bamzooki,: Wikipedia http://en.wikipedia.org/wiki/Bamzooki. Accessed 14 Apr 2014
4. Wikipedia, Bernoulli's principle, http://en.wikipedia.org/wiki/Bernoulli's_principle. Accessed 14 Apr 2014
5. Vogel, S., Ellington, C.P., Kilgore, D.C.: Life in moving fluids. The physical biology of flow. Princeton University press, Princeton (1973)
6. Hoogland, J.L.: Black-Tailed Prairie Dog: social life of a burrowing mammal. University of Chicago Press, Chicago (1995)
7. Ask nature. www.asknature.org/strategy/e27b89ebcdec8c9b5b2cd9ac84b8f8a0#.U0tDrlWS yo0. Accessed 14 Apr 2014
8. Like a fish technologies. www.likeafish.biz. Accessed 14 Apr 2014
9. Deep sea news. http://deepseanews.com/2014/01/triton-not-dive-or-dive-not-there-is-no-triton/. Accessed 14 Apr 2012
10. Straus, M.B., Askenov, I.V.: Diving Science, essential physiology and medicine for divers. Human Kinetics. II. USA (2004)
11. Benyus, J.M.: Biomimicry in action, (TEDGlobal 2009), http://www.ted.com/talks/janine_benyus_biomimicry_in_action
12. Connor, S.: Plan to avert global warming by cooling planet artificially could cause climate chaos, The Independent, 8 Jan 2014
13. Savory, A.: What Could the Massacre of 40,000 Elephants Possibly Teach Us?. http://www.ted.com/talks/allan_savory_how_to_green_the_world_s_deserts_and_reverse_climate_change (TED 2013). Accessed 14 Apr 2012
14. Crawford, E.K., Worsham, N.L., Swinehart, E.R.: Benefits derived from companion animals, and the use of the term "attachment". Anthrozoos: Multidiscip. J. Interact. People Anim. **19**(2), 98–112(15) (2006). (Bloomsbury Journals (formerly Berg Journals))
15. 3D printing. Wikipedia, http://en.wikipedia.org/wiki/3D_printing. Accessed 14 Apr 2014
16. Jonkers, H.M.: PDF Bacteria-based self-healing concrete, Delft University of Technology, Faculty of Civil Engineering and Geosciences, Department of Materials and Environment–Microlab, Delft, the Netherlands (2011)
17. Koh, E., Kim, N.-K., Shin, J., Kim, Y.-W.: Polyurethane microcapsules for self-healing paint coatings. RSC, 31 (2014). Accessed 12 Mar 2014
18. US ONR develops self-healing anti-corrosion paint for military vehicles, 20 Mar 2014 naval-technology.com www.naval-technology.com/news/newsus-onr-develops-self-healing-anti-corrosion-paint-for-military-vehicles-4200285. Accessed 14 Apr 2014
19. Cocoon_FS: Pohl Architects Unveils Prefab Plankton-Inspired Pod Building in Germany, by Allison Leahy, 01/31/12, Inhabitat-Sustainable Design Innovation, Eco Architecture, Green Building http://inhabitat.com/cocoon_fs-pohl-architects-unveils-prefab-plankton-inspired-pod-building-in-germany/. Accessed 14 Apr 2014

20. Bottlenecks and Founder Effects, http://evolution.berkeley.edu/evosite/evo101/IIID3Bottle necks.shtml. Accessed 14 Apr 2014
21. Navier Stokes equations, Wikipedia http://en.wikipedia.org/wiki/Navier%E2%80%93Stokes_equations. Accessed 14 Apr 2014
22. Bradley, L.: Chaos and Fractals, Strange Attractors, (2010) http://www.stsci.edu/~lbradley/seminar/attractors.html. Accessed 14 Apr 2014
23. Sample, I.: Warming hits tipping point,, The guardian, Guardian unlimited. http://xxx.biologicaldiversity.org/news/media-archive/Warming%20Hits%20Tipping%20Point.pdf. Accessed 15 Apr 2014
24. Lenton, T.M., Ciscar, J-C.: Integrating tipping points into climate impact Assessments. Clim. Change. *117*(3): 585–597. Springer, Published online (2013). Accessed 29 Aug 2012
25. Talk, Predicting Climate in a Chaotic World: How Certain Can We Be? Professor Timothy Palmer New England aquarium, November 1, 2012, Lorenz Centre, MIT Earth, Atmospheric and earth sciences http://web.mit.edu/lorenzcenter/activities/past-events.html. Accessed 15 Apr 2012
26. Shutdown of thermohaline circulation, Wikipedia, http://en.wikipedia.org/wiki/Shutdown_of_thermohaline_circulation. Accessed 15 Apr 2012
27. Navas, C.A., Carvalho, J.E.: Aestivation: molecular and physiological aspects. Spinger, Berlin (2010)
28. True mammalian Hibernation. http://www.britannica.com/EBchecked/topic/169514/dormancy/48538/True-mammalian-hibernation. Accessed 15 Apr 2012
29. Tisherman, S.A., Rodriguez, A., Safar, P.: Therapeutic hypothermia in Traumatology, Surg. Clin. N. Am. *796*, 1269–1289, Elsevier, http://linkinghub.elsevier.com/retrieve/pii/S0039610905700773?via=sd (2013). Accessed 15 Apr 2014
30. Issels M.I.: Fever Therapy: restoring regulatory mechanisms a powerful immune enhancement an overview. http://www.issels.com/publications/FeverTherapy.aspx#sthash.sVwAnB7y.711dK6UQ.dpbs (2002). Accessed 15 Apr 2014
31. Holick, M.F., Chen, T.C.: Vitamin D deficiency: a worldwide problem with health consequences. Am. J. Clin. Nutr. **874**, http://ajcn.nutrition.org/content/87/4/1080S.short (2008). Accessed 14 Apr 2014
32. Rayleigh scattering. Encyclopædia Britannica. Encyclopædia Britannica Online (2007). Accessed 16 Nov 2014
33. Thapan, K., Arendt, J., Skene, D.J.: An action spectrum for melatonin suppression: evidence for a novel non-rod, non-cone photoreceptor system in humans, J. Phys. **535**, 261–267. Accessed 15 Aug 2001
34. Blue light has a dark side, Harvard Health Publications, Boston http://www.health.harvard.edu/newsletters/Harvard_Health_Letter/2012/May/blue-light-has-a-dark-side/ (2012). Accessed 14 Apr 2014
35. Wright, H.R., Lack, L.C.: Effect of light wavelength on suppression and phase delay of the melatonin rhythm, Chronobiol_Int. (5): 801-8. Accessed 18 Sep 2001
36. Parry, G.: Which of our basic physical assumptions are wrong?, http://www.fqxi.org/data/essay-contestfiles/Parry_PARRY_FQXi_competitio.pdf (2012). Accessed 15 Apr 2014
37. Sagan, C.: Pale Blue Dot, A Vision of the Human Future in Space, Random House, New York (1994)

Appendix
List of Winners

First Prize
Sabine Hossenfelder: *How to Save the World*

Second Prizes
Daniel Dewey: *Crucial Phenomena*

Jens Niemeyer: *How to Avoid Steering Blindly: The Case for a Robust Repository of Human Knowledge*

Third Prizes
Preston Estep III and Alexander Hoekstra: *The Leverage and Centrality of Mind*

Mohammed Khalil: *Improving Science for a Better Future*

Travis Norsen: *Back to the Future: Crowdsourcing Innovation by Refocusing Science Education*

Dean Rickles: *A Participatory Future of Humanity*

Rick Searle: *The Cartography of the Future: Recovering Utopia for the 21st Century*

Tejinder Singh: *Enlightenment is not for the Buddha Alone*

Fourth Prizes
Tommaso Bolognesi: *Humanity is Much More Than the Sum of Humans*

Jonathan Dickau: *Recognizing the Value of Play*

From the Foundational Questions Institute website: http://fqxi.org/community/essay/winners/2014.1.

George Gantz: *The Tip of the Spear*

Flavio Mercati: *U-turn or u die*

Georgina Parry: *Smooth Seas Do Not Make Good Sailors*

Titles in this Series

© Springer International Publishing Switzerland 2016
A. Aguirre et al. (eds.), *How Should Humanity Steer the Future?*,
The Frontiers Collection, DOI 10.1007/978-3-319-20717-9

Extreme States of Matter
on Earth and in the Cosmos
By Vladimir E. Fortov

Searching for Extraterrestrial Intelligence
SETI Past, Present, and Future
Ed. by H. Paul Shuch

Essential Building Blocks of Human Nature
Ed. by Ulrich J. Frey, Charlotte Störmer and Kai P. Willführ

Mindful Universe
Quantum Mechanics and the Participating Observer
By Henry P. Stapp

Principles of Evolution
From the Planck Epoch to Complex Multicellular Life
Ed. by Hildegard Meyer-Ortmanns and Stefan Thurner

The Second Law of Economics
Energy, Entropy, and the Origins of Wealth
By Reiner Kümmel

States of Consciousness
Experimental Insights into Meditation, Waking, Sleep and Dreams
Ed. by Dean Cvetkovic and Irena Cosic

Elegance and Enigma
The Quantum Interviews
Ed. by Maximilian Schlosshauer

Humans on Earth
From Origins to Possible Futures
By Filipe Duarte Santos

Evolution 2.0
Implications of Darwinism in Philosophy and the Social and Natural Sciences
Ed. by Martin Brinkworth and Friedel Weinert

Probability in Physics
Ed. by Yemima Ben-Menahem and Meir Hemmo

Chips 2020
A Guide to the Future of Nanoelectronics
Ed. by Bernd Hoefflinger

From the Web to the Grid and Beyond
Computing Paradigms Driven by High-Energy Physics
Ed. by René Brun, Federico Carminati and Giuliana Galli Carminati

Printed in the United States
By Bookmasters